The Great Books 2 Reading & Discussion Program

FIRST SERIES · VOLUME ONE

The Great Books Foundation

A Nonprofit Educational Corporation

Designed by Don Walkoe Design, Chicago

Handmade marbled paper, photographed on cover,
courtesy of Andrews, Nelson, Whitehead.

ISBN 0-945159-76-5

9 8 7 6 5

Published and distributed by

The Great Books Foundation
A Nonprofit Educational Corporation
35 East Wacker Drive, Suite 2300
Chicago, Illinois 60601-2298

Acknowledgments

"Rothschild's Fiddle" from *The Oxford Chekhov, Vol. VII: Stories 1893–1895*, translated and edited by Ronald Hingley. Copyright 1978 by Ronald Hingley. Reprinted by permission of the publisher, Oxford University Press.

"On Happiness" from *Nichomachean Ethics* by Aristotle, translated by Martin Ostwald. Copyright 1962 by The Bobbs-Merrill Company, Inc. Reprinted by permission of the publisher, The Bobbs-Merrill Company, Inc.

"The Apology" by Plato from *Socrates and Legal Obligation*, translated by R. E. Allen. Copyright 1980 by the University of Minnesota. Reprinted by permission of the publisher, The University of Minnesota Press.

"Conscience" from *Lectures on Ethics* by Immanuel Kant, translated by Louis Infield. Published by Methuen & Co., Ltd. Reprinted by permission of Associated Book Publishers, Ltd.

"Alienated Labour" from *Karl Marx, Early Writings*, translated and edited by T. B. Bottomore. Copyright 1963 by T. B. Bottomore. Reprinted by permission of the publisher, McGraw-Hill Book Company.

"Civilization and Its Discontents" from *Civilization and Its Discontents*, translated by James Strachey. Copyright 1961 by James Strachey. Reprinted by permission of the publisher, W. W. Norton & Company, Inc.

"The Social Contract" from *On the Social Contract* by Jean-Jacques Rousseau, translated by Judith R. Masters, edited by Roger D. Masters. Copyright 1978 by St. Martin's Press, Inc. Reprinted by permission of the publisher, St. Martin's Press, Inc.

"Of Justice and Injustice" from *Hume's Moral and Political Philosophy*, edited by Henry D. Aiken. Copyright 1948 by Hafner Press, a division of Macmillan Publishing Company, Inc. Reprinted by permission of the publisher, Macmillan Publishing Company, Inc.

"Individual Freedom" from *The Philosophy of Money* by Georg Simmel, translated by Tom Bottomore and David Frisby. Copyright 1978 by Routledge and Kegan Paul, Ltd. Reprinted by permission of the publisher, Routledge and Kegan Paul, Ltd.

"Antigone" from *The Antigone of Sophocles: An English Version* by Dudley Fitts and Robert Fitzgerald. Copyright 1967 by Dudley Fitts and Robert Fitzgerald. Reprinted by permission of the publisher, Harcourt Brace Jovanovich, Inc.

CONTENTS

*

A source note appears, together with biographical information about the author, opposite the opening page of each work in this series. Footnotes by the author are not bracketed; footnotes by GBF or a translator are [bracketed].

•

ANTON PAVLOVICH CHEKHOV was born in 1860 in provincial Taganrog, Russia to a shopkeeper who had been born a serf. Chekhov was educated at a local school run by Greeks. When he was seventeen, his father's business went bankrupt, and the family moved to Moscow except for Chekhov, who remained in Taganrog to finish school. He had already begun writing. His humorous sketches, stories, and reportage appeared under numerous pseudonyms—"Blockhead," "Starling," "A Prosaic Poet," etc.—in small newspapers and magazines. In 1879 Chekhov joined his family in Moscow and entered the University of Moscow to study medicine, meanwhile writing cartoon captions, calendars, parodies, advertisements, a gossip column, theater reviews, and a detective novel, as well as short fiction. He graduated as a doctor in 1884 and practiced medicine. Chekhov's first book was a collection of short stories, *Motley Stories* (1886). He had written most of his stories by the age of thirty, and then turned to the writing of plays. Chekhov died of tuberculosis in 1904.

From *The Oxford Chekhov, Volume VII: Stories 1893–1895*, translated and edited by Ronald Hingley. Publisher: Oxford University Press, 1978.

Rothschild's Fiddle

It was a small town, more wretched than a village, and almost all the inhabitants were old folk with a depressingly low death rate. Nor were many coffins required at the hospital and jail. In a word, business was bad. Had Jacob Ivanov been making coffins in a county town he would probably have owned a house and been called "mister." But in this dump he was plain Jacob, his street nickname was "Bronze" for whatever reason, and he lived as miserably as any farm laborer in his little old one-roomed shack which housed himself, his Martha, a stove, a double bed, coffins, his workbench and all his household goods.

Jacob made good solid coffins. For men—village and working-class folk—he made them to his own height, and never got them wrong because he was taller and stronger than anyone, even in the jail, though now seventy years old. For the gentry, though, and for women he made them to measure, using an iron ruler. He was not at all keen on orders for children's coffins, which he would knock up contemptuously without measuring. And when paid for them he would say that he "quite frankly set no store by such trifles."

His fiddle brought him a small income on top of his trade. A Jewish band usually played at weddings in the town, conducted by the tinker Moses Shakhkes who took more than half the proceeds. And since Jacob was a fine fiddler, especially with Russian folk tunes, Shakhkes sometimes asked him to join the band for fifty kopeks a day plus tips. Straightaway it made his

1

face sweat and turn crimson, did sitting in the band. It was hot, there was a stifling smell of garlic, his fiddle squeaked. By his right ear wheezed the double-bass, by his left sobbed the flute played by a red-haired, emaciated Jew with a network of red and blue veins on his face. He was known as Rothschild after the noted millionaire. Now, this bloody little Jew even contrived to play the merriest tunes in lachrymose style. For no obvious reason Jacob became more and more obsessed by hatred and contempt for Jews, and for Rothschild in particular. He started picking on him and swearing at him. Once he made to beat him, whereat Rothschild took umbrage.

"I respect your talent, otherwise I am long ago throwing you out of window," he said with an enraged glare.

Then he burst into tears. This was why Bronze wasn't often asked to play in the band, but only in some dire crisis, when one of the Jews was unavailable.

Jacob was always in a bad mood because of the appalling waste of money he had to endure. For instance, it was a sin to work on a Sunday or a Saint's Day, while Mondays were unlucky, so that made two hundred odd days a year when you had to sit around idle. And that was all so much money wasted. If someone in town held a wedding without music, or Shakhkes didn't ask Jacob to play, that meant still more losses. The police superintendent had been ill for two years now. He was wasting away, and Jacob had waited impatiently for him to die, but the man had left for treatment in the county town, and damned if he didn't peg out there. Now, that was at least ten rubles down the drain, as his would have been an expensive coffin complete with brocade lining. Thoughts of these losses hounded Jacob mostly at night. He would put his fiddle on the bed beside him, and when some such tomfoolery preyed on his mind he would touch the strings and the fiddle would twang in the darkness. That made him feel better.

On the sixth of May in the previous year Martha had suddenly fallen ill. The old woman breathed heavily, drank a lot of water, was unsteady on her feet, but she would still do the stove herself of a morning, and even fetch the water. By evening, though, she would already be in bed. Jacob fiddled away all day. But when it was quite dark he took the book in which he listed his losses daily and began, out of sheer boredom, to add up the annual total. It came to more than a thousand rubles. This so shocked him that he flung his abacus on the floor and stamped his feet. Then he picked up the abacus and clicked away again for a while, sighing deep, heartfelt sighs. His face was purple and wet with sweat. He was thinking that if he had put that lost thousand in the bank he would have received at least forty rubles' interest a year. So that was forty more rubles down the drain. However hard you tried to wriggle out of it, everything was just a dead loss in fact.

Then he suddenly heard Martha call out, "Jacob, I'm dying."

He looked round at his wife. Her face was flushed in the heat, her expression was exceptionally bright and joyous. Accustomed to her pale face and timid, unhappy expression, Bronze was put out. She really did look as if she was dying, glad to be saying a permanent good-bye to hut, coffins and Jacob at long last.

Gazing at the ceiling and moving her lips, she looked happy, as if she could actually see her savior Death and was whispering to him.

It was dawn and the first rays were seen through the window. As he looked at the old woman, it vaguely occurred to Jacob that for some reason he had never shown her any affection all his life. Never had he been kind to her, never had he thought of buying her a kerchief or bringing her sweetmeats from a wedding. All he had done was yell at her, blame her for his "losses," threaten to punch her. True, he never had hit her. Still he had frightened her, she had always been petrified with fear.

Yes, he had said she couldn't have tea because they had enough other expenses without that, so she only drank hot water. And now he knew why she looked so strangely joyous, and a chill went through him.

When it was fully light he borrowed a neighbor's horse and drove Martha to hospital. There were not many patients, so he did not have long to wait. Only about three hours. To his great joy the patients were not received on this occasion by the doctor, who was ill himself, but by his assistant Maxim, an old fellow said by everyone in town to be better than the doctor, drunken brawler though he was.

"I humbly greet you," said Jacob, taking his old woman into the consulting room. "You must excuse us troubling you with our trifling affairs, sir. Now, as you see, guv'nor, my old woman has fallen sick. She's my better half in a manner of speaking, if you'll pardon the expression—"

Frowning, stroking his side-whiskers, the white-eyebrowed assistant examined the old woman, who sat hunched on a stool, wizened, sharp-nosed, open-mouthed, her profile like a thirsty bird's.

"Hurrumph. Well, yes," the assistant slowly pronounced, and sighed. "It's influenza, fever perhaps. There's typhus in town. Ah well, the old woman's lived her life, praise the Lord. How old is she?"

"Seventy come next year, guv'nor."

"Ah well, her life's over. Time she was on her way."

"It's true enough, what you just said, sir." Jacob smiled out of politeness. "And we thanks you most kindly for being so nice about it, like. But, if you'll pardon the expression, every insect wants to live."

"Not half it does." The assistant's tone suggested that it depended on him whether the old woman lived or died. "Now then, my good man, you put a cold compress on her head and give her one of these powders twice a day. And now cheerio to you. A very bong jour."

From his face Jacob could tell that it was all up, and that no powders would help now. Obviously Martha was going to die soon, either today or tomorrow. He gave the assistant's elbow a push and winked.

"We ought to cup her, Mr. Maxim sir," he said in a low voice.

"Haven't the time, my good man. Take your old woman and be off with you. So long and all that."

"Begging your kindness, sir," implored Jacob. "As you know, mister, if it was her guts or her innards, like, what was sick, then it's powders and drops she should have. But this here is a chill, and the great thing with chills is to bleed 'em, sir."

But the assistant had already called for his next patient, and a village woman with a little boy had come into the consulting room.

"Buzz off you, beat it!" The assistant frowned at Jacob. "Don't hang around."

"Then at least put some leeches to her. I'll be grateful to you all my life, I will."

The assistant lost his temper.

"Don't you bandy words with me," he yelled. "D-damned oaf!"

Jacob lost his temper too, and turned completely crimson. But he grabbed Martha's arm without a single word and took her out of the room. Only when they were getting into their cart did he cast a stern, mocking look at that hospital.

"They're a high and mighty lot round here," he said. "He'd have cupped a rich man, I'll be bound, but for a poor one he grudges even a single leech. Bastards!"

They arrived home, and Martha, after entering the house, stood for about ten minutes gripping the stove. If she was to lie down Jacob would talk about all the money he'd lost and blame her for lolling about and not wanting to work—or so she thought. And Jacob looked at her miserably, remembering that tomorrow was St. John's Day, and the day after that was St. Nicholas' Day, after which came Sunday and Unlucky Monday. That made four days when he couldn't work. But Martha was sure to die on one of those days, so he must make the coffin today. He took his iron ruler, went up to the old woman and measured her. Then she lay down and he crossed himself and started on the coffin.

When the work was finished, Jacob put on his spectacles and wrote in his book.

"Martha Ivanov: to one coffin, 2 rubles 40 kopeks."

He sighed. The old woman lay there all the time silently, her eyes shut, but when it grew dark that evening she suddenly called the old man.

"Remember fifty years ago, Jacob?" She looked at him happily. "God gave us a little fair-haired baby, remember? We were always sitting by the river, you and I, singing songs under the willow tree." She laughed bitterly. "The little girl died."

Jacob cudgeled his brains, but could recall neither baby nor willow. "You're imagining things."

The priest came and gave the last rites, whereupon Martha mumbled something or other. By morning she was gone.

Old women neighbors washed her, dressed her, laid her in her coffin. So as not to waste money on the sexton, Jacob read the lesson himself, and he got the grave for nothing because the cemetery caretaker was a crony of his. Four peasants bore the coffin to the cemetery out of respect, not for money. It was followed by old women, beggars and two village idiots while people in the street crossed themselves piously. Jacob was delighted that it was all so right and seemly, that it didn't cost much or hurt anyone's feelings. As he said good-bye to Martha for the last time he touched the coffin.

"Good workmanship, that," he thought.

But on his way back from the cemetery he was overcome by a great sorrow. He felt vaguely unwell. His breath came hot and heavy, his legs were weak, he felt thirsty. Then various thoughts began to prey on his mind. He again remembered that never in his life had he been kind to Martha or shown her affection. The fifty-two years of their life together in one hut— it seemed such a long, long time. But somehow he had never given her a thought in all that time, he had no more noticed her than a cat or dog. But she had made up the stove every day, hadn't she? She had cooked, baked, fetched water, cut wood, shared his bed. And when he came back from weddings drunk she would reverently hang his fiddle on the wall and put him to bed—all this in silence, looking scared and troubled.

Rothschild approached Jacob, smiling and bowing.

"I been looking for you, mister," he said. "Mister Moses sends his respects, says he vonts you at once."

Jacob wasn't interested. He wanted to cry.

"Leave me alone." He walked on.

"Vot are you doing?" Rothschild ran ahead, much alarmed. "Mister Moses'll be offended. You're to come at once, said he."

Out of breath, blinking, with all those red freckles, the Jew disgusted Jacob. The green frock coat with the black patches, his whole frail, puny figure—what a loathsome sight.

"Keep out of my way, Garlic-breath," shouted Jacob. "You leave me alone."

The Jew, angered, also shouted. "You are being quiet please or I am throwing you over fence."

"Out of my sight, you!" bellowed Jacob, pouncing on him with clenched fists. "Proper poison, them greasy bastards are."

Scared to death, Rothschild crouched down, waving his hands above his head as if warding off blows, then jumped up and scampered off as fast as he could, hopping about and flapping his arms as he ran. You could see the quaking of his long, thin back. At this the street urchins gleefully rushed after him shouting "Dirty Yid!" Barking dogs chased him too. Someone roared with laughter and then whistled, the dogs barked louder and in closer harmony.

Then a dog must have bitten Rothschild, for a shout of pain and despair was heard.

Jacob walked on the common, then started off along the edge of the town without knowing where he was going. "There's old Jake, there he goes," shouted the boys. Then he came to the river. Here sandpipers swooped and twittered, ducks quacked. The sun's heat beat down and the water sparkled till it hurt the eyes. Walking along the towpath, Jacob saw a buxom, red-cheeked woman emerge from a bathing hut.

"Damn performing seal," he thought.

Not far from the bathing hut boys were fishing for crayfish, using meat for bait. They saw him.

"Hey, there's old Jake," they shouted nastily.

Then came the broad old willow tree with its huge hollow and crows' nests.

Suddenly Jacob's memory threw up a vivid image of that fair-haired baby and the willow that Martha had spoken of. Yes, it was the same willow—so green, so quiet, so sad.

How old it had grown, poor thing.

He sat beneath it and began remembering. On the other bank, now a water meadow, had been a silver-birch forest, and over there on that bare hill on the horizon the dark blue bulk of an ancient pine wood. Barges had plied up and down the river. But now it was all flat and bare with the one little silver birch on the near side, slim and youthful as a young girl. There were only ducks and geese on the river, and it was hard to think that barges had ever passed here. Even the geese seemed fewer. Jacob shut his eyes and pictured vast flocks of white geese swooping towards each other.

How was it, he wondered, that he had never been by the river in the last forty or fifty years of his life. Or, if he had, it had made no impression on him. Why, this was a proper river, not just any old stream. You could fish it, you could sell the fish to shopkeepers, clerks and the man who kept the station bar, you could put the money in the bank. You could sail a boat from one riverside estate to another playing your fiddle, and all manner of folk would pay you for it. You could try starting up the barges again—better than making coffins, that was. Then you could breed geese, slaughter them and send them to Moscow in winter. "The down alone would fetch ten rubles a year, I'll be bound." But he had let all this go by, he had done nothing about it. Oh, what a waste, what a waste of money! If you put it all together—fishing, fiddling, barging, goose-slaughtering—what a lot of money you'd have made. But none of it had happened, not even in your dreams. Life had flowed past without profit, without enjoyment—gone aimlessly, leaving nothing to show for it. The future was empty. And if you looked back there was only all the awful waste of money that sent shivers down your spine. Why couldn't a man live without all that loss and waste? And why, he wondered, had they cut down the birch forest? And the pine wood? Why wasn't that common put to use? Why do people always do the wrong things? Why had Jacob spent all his life cursing, bellowing, threatening people with his fists, ill-treating his wife? And what, oh what, was the point of scaring and insulting that Jew just now? Why are people generally such a nuisance to each other? After all, it's all such a waste of money, a terrible waste it is. Without the hate and malice folks could get a lot of profit out of each other.

That evening and night he had visions of baby, willow, fish, dead geese, of Martha with her thirsty bird's profile, and of Rothschild's wretched, pale face, while various other gargoyle-like faces advanced on him from all sides muttering about all the waste of money. He tossed and turned, he got out of bed half a dozen times to play his fiddle.

Next morning he forced himself to get up and went to the hospital. That same Maxim told him to put a cold compress on his head and gave him powders, but his look and tone made Jacob realize that it was all up and that no powders would help now. Later, on his way home, he reckoned that death would be pure gain to him. He wouldn't have to eat, drink, pay taxes or offend folk. And since a man lies in his grave not just one but hundreds and thousands of years, the profit would be colossal. Man's life is debit, his death credit. The argument was correct, of course, but painfully disagreeable too. Why are things so oddly arranged? You only live once, so why don't you get anything out of it?

He didn't mind dying, but when he got home and saw his fiddle his heart missed a beat and he felt sorry. He couldn't take his fiddle with him to the grave, so it would be orphaned and go the way of the birches and the pines. Nothing in this world has ever come to anything, nothing ever will. Jacob went out of the hut and sat in the doorway clasping the fiddle to his breast. Thinking of his wasted, profitless life, he started playing he knew not what, but it came out poignantly moving and tears coursed down his cheeks. The harder he thought the sadder grew the fiddle's song.

The latch squeaked twice and Rothschild appeared at the garden gate. He crossed half the yard boldly, but when he saw Jacob he suddenly stopped, cringed and—through fear, no doubt—gesticulated as if trying to indicate the time with his fingers.

"Come along then," said Jacob kindly, beckoning him. "It's all right."

Looking at him mistrustfully and fearfully, Rothschild began to approach but stopped a few feet away.

"Don't hit me, I beg you." He squatted down. "It's Mister Moses has sent me again. Never fear, says he, you go to Jacob again—tell him we can't do without him, he says. There's a vedding on Vednesday. Aye, that there is. Mister Shapovalov is marrying his daughter to a fine young man. A rich folks' vedding this, and no mistake!" The Jew screwed up one eye.

"Can't be done." Jacob breathed heavily. "I'm ill, son."

He again struck up, his tears spurting on to the fiddle. Rothschild listened carefully, standing sideways on, arms crossed on his breast. His scared, baffled look gradually gave way to a sorrowful, suffering expression. He rolled his eyes as if in anguished delight.

"A-a-ah!" he said as the tears crawled down his cheeks and splashed on his green frock coat.

After that Jacob lay down all day, sick at heart. When the priest heard his confession that evening and asked whether he remembered committing any particular sin he exerted his failing memory and once more recalled Martha's unhappy face and the desperate yell of the Jew bitten by a dog.

"Give my fiddle to Rothschild," he said in a voice barely audible.

"Very well," the priest answered.

Now everyone in town wants to know where Rothschild got such a fine fiddle. Did he buy it, did he steal it? Or did someone leave it with him as a pledge? He only plays the fiddle now, having given up the flute long ago. From his bow there flow those same poignant strains which used to come from his flute. But when he tries to repeat the tune Jacob had played in his doorway the outcome is so sad and mournful that his listeners weep and he ends by rolling his eyes up with an "A-a-ah!"

So popular is this new tune in town that merchants and officials are always asking Rothschild over and making him play it a dozen times.

ARISTOTLE was born in Stagirus, a Greek colony in Macedonia, in 384 B.C. His father was physician to the Macedonian court, and Aristotle probably studied biology and medicine as a boy. At seventeen, he entered Plato's Academy in Athens, where he studied and lectured for twenty years. When Plato died, Aristotle left Athens and taught at branches of the Academy in Asia Minor. In 343 he was invited to serve as a tutor to thirteen-year-old Alexander, future king of Macedonia. Aristotle's treatises were most likely presented first as lectures or reading assignments for his students in the school he opened in Athens after tutoring Alexander. The treatises include *The Physics, On the Heavens, De Anima, The Nichomachean Ethics, the Politics,* and *The Poetics.* In 323 Aristotle's friendships with Macedonians made him a target of anti-Macedonian feeling then current in Athens. He left Athens and died in Chalcis in 322 B.C.

From *Nichomachean Ethics,* translated by Martin Ostwald. Publisher: Bobbs-Merrill Co., Inc., 1962. Book I, pages 3–9 and 14–32; and Book X, pages 288–91.

On Happiness

The good as the aim of action

Every art or applied science and every systematic investigation, and similarly every action and choice, seem to aim at some good; the good, therefore, has been well defined as that at which all things aim. But it is clear that there is a difference in the ends at which they aim: in some cases the activity is the end, in others the end is some product beyond the activity. In cases where the end lies beyond the action the product is naturally superior to the activity.

Since there are many activities, arts, and sciences, the number of ends is correspondingly large: of medicine the end is health, of shipbuilding a vessel, of strategy, victory, and of household management, wealth. In many instances several such pursuits are grouped together under a single capacity: the art of bridle-making, for example, and everything else pertaining to the equipment of a horse are grouped together under horsemanship; horsemanship in turn, along with every other military action, is grouped together under strategy; and other pursuits are grouped together under other capacities. In all these cases the ends of the master sciences are preferable to the ends of the subordinate sciences, since the latter are pursued for the sake of the former. This is true whether the ends of the actions lie in the activities themselves or, as is the case in the disciplines just mentioned, in something beyond the activities.

Politics as the master science of the good

Now, if there exists an end in the realm of action which we desire for its own sake, an end which determines all our other desires; if, in other words, we do not make all our choices for the sake of something else—for in this way the process will go on infinitely so that our desire would be futile and pointless—then obviously this end will be the good, that is, the highest good. Will not the knowledge of this good, consequently, be very important to our lives? Would it not better equip us, like archers who have a target to aim at, to hit the proper mark? If so, we must try to comprehend in outline at least what this good is and to which branch of knowledge or to which capacity it belongs.

This good, one should think, belongs to the most sovereign and most comprehensive master science, and politics[1] clearly fits this description. For it determines which sciences ought to exist in states, what kind of sciences each group of citizens must learn, and what degree of proficiency each must attain. We observe further that the most honored capacities, such as strategy, household management, and oratory, are contained in politics. Since this science uses the rest of the sciences, and since, moreover, it legislates what people are to do and what they are not to do, its end seems to embrace the ends of the other sciences. Thus it follows that the end of politics is the good for man. For even if the good is the same for the individual and the state, the good of the state clearly is the greater and more perfect thing to attain and to safeguard. The attainment of the good for one

---

[1] [*Politikē* is the science of the city-state, the *polis,* and its members, not merely in our narrow "political" sense of the word but also in the sense that a civilized human existence is, according to Plato and Aristotle, only possible in the *polis.* Thus *politikē* involves not only the science of the state, "politics," but of our concept of "society" as well. — TRANS.]

man alone is, to be sure, a source of satisfaction; yet to secure it for a nation and for states is nobler and more divine. In short, these are the aims of our investigation, which is in a sense an investigation of social and political matters.

The limitations of ethics and politics

Our discussion will be adequate if it achieves clarity within the limits of the subject matter. For precision cannot be expected in the treatment of all subjects alike, any more than it can be expected in all manufactured articles. Problems of what is noble and just, which politics examines, present so much variety and irregularity that some people believe that they exist only by convention and not by nature. The problem of the good, too, presents a similar kind of irregularity, because in many cases good things bring harmful results. There are instances of men ruined by wealth, and others by courage. Therefore, in a discussion of such subjects, which has to start from a basis of this kind, we must be satisfied to indicate the truth with a rough and general sketch: when the subject and the basis of a discussion consist of matters that hold good only as a general rule, but not always, the conclusions reached must be of the same order. The various points that are made must be received in the same spirit. For a well-schooled man is one who searches for that degree of precision in each kind of study which the nature of the subject at hand admits: it is obviously just as foolish to accept arguments of probability from a mathematician as to demand strict demonstrations from an orator.

Each man can judge competently the things he knows, and of these he is a good judge. Accordingly, a good judge in each particular field is one who has been trained in it, and a good judge in general, a man who has received an all-round schooling. For that reason, a young man is not equipped to be a student of politics; for he has no experience in the actions which life demands of him, and these actions form the basis and subject matter of the discussion. Moreover, since he follows his emotions, his study will be pointless and unprofitable, for the end of this kind of study is not knowledge but action. Whether he is young in years or immature in character makes no difference; for his deficiency is not a matter of time but of living and of pursuing all his interests under the influence of his emotions. Knowledge brings no benefit to this kind of person, just as it brings none to the morally weak. But those who regulate their desires and actions by a rational principle will greatly benefit from a knowledge of this subject. So much by way of a preface about the student, the limitations which have to be accepted, and the objective before us.

Happiness is the good, but many views are held about it

To resume the discussion: since all knowledge and every choice is directed toward some good, let us discuss what is in our view the aim of politics, i.e., the highest good attainable by action. As far as its name is concerned, most people would probably agree: for both the common run of people and cultivated men call it happiness, and understand by "being happy" the same as "living well" and "doing well." But when it comes to defining what happiness is, they disagree, and the account given by the common run differs from that of the philosophers. The former say it is some clear and obvious good, such as pleasure, wealth, or honor; some say it is one thing and others another, and often the very same person identifies it with different things at different times: when he is sick he thinks it is health, and when he is poor he says it is wealth; and when people are conscious of their own ignorance, they admire those who talk above their heads in accents of greatness. Some thinkers used to believe that there exists over and above these many goods another good, good in itself and by itself, which also is the cause of good in all these things. An examination of all the different opinions would perhaps be a little pointless, and it is sufficient to concentrate on those which are most in evidence or which seem to make some sort of sense. . . .

Various views on the highest good

. . . It is not unreasonable that men should derive their concept of the good and of happiness from the lives which they lead. The common run of people and the most vulgar identify it with pleasure, and for that reason are satisfied with a life of enjoyment. For the most notable kinds of life are three: the life just mentioned, the political life, and the contemplative life.

The common run of people, as we saw, betray their utter slavishness in their preference for a life suitable to cattle; but their views seem plausible because many people in high places share the feelings of Sardanapallus.[2] Cultivated and active men, on the other hand, believe the good to be honor, for honor, one might say, is the end of the political life. But this is clearly too superficial an answer: for honor seems to depend on those who confer it rather than on him who receives it, whereas our guess is that the good is a man's own possession which cannot easily be taken away from him. Furthermore, men seem to pursue honor to assure themselves of their own worth; at any rate, they seek to be honored by sensible men and by those who know them, and they want to be honored on the basis of their virtue or excellence. Obviously, then, excellence, as far as they are concerned, is better than honor. One might perhaps even go so far as to consider excellence rather than honor as the end of political life. However, even excellence proves to be imperfect

---

[2] [Sardanapallus is the Hellenized name of the Assyrian king Ashurbanipal (669–626 B.C.). Many stories about his sensual excesses were current in antiquity. — TRANS.]

as an end: for a man might possibly possess it while asleep or while being inactive all his life, and while, in addition, undergoing the greatest suffering and misfortune. Nobody would call the life of such a man happy, except for the sake of maintaining an argument. . . . In the third place there is the contemplative life, which we shall examine later on. As for the money-maker, his life is led under some kind of constraint: clearly, wealth is not the good which we are trying to find, for it is only useful, i.e., it is a means to something else. Hence one might rather regard the aforementioned objects as ends, since they are valued for their own sake. But even they prove not to be the good, though many words have been wasted to show that they are. Accordingly, we may dismiss them. . . .

The good is final and self-sufficient; happiness is defined

Let us return again to our investigation into the nature of the good which we are seeking. It is evidently something different in different actions and in each art: it is one thing in medicine, another in strategy, and another again in each of the other arts. What, then, is the good of each? Is it not that for the sake of which everything else is done? That means it is health in the case of medicine, victory in the case of strategy, a house in the case of building, a different thing in the case of different arts, and in all actions and choices it is the end. For it is for the sake of the end that all else is done. Thus, if there is some one end for all that we do, this would be the good attainable by action; if there are several ends, they will be the goods attainable by action.

. . . Since there are evidently several ends, and since we choose some of these—e.g., wealth, flutes, and instruments generally— as a means to something else, it is obvious that not all ends are final. The highest good, on the other hand, must be something final. Thus, if there is only one final end, this will be the good we are seeking; if there are several, it will be the most final and perfect of them. We call that which is pursued as an end in itself more final than an end which is pursued for the sake of something else; and what is never chosen as a means to something else we call more final than that which is chosen both as an end in itself and as a means to something else. What is always chosen as an end in itself and never as a means to something else is called final in an unqualified sense. This description seems to apply to happiness above all else: for we always choose happiness as an end in itself and never for the sake of something else. Honor, pleasure, intelligence, and all virtue we choose partly for themselves—for we would choose each of them even if no further advantage would accrue from them—but we also choose them partly for the sake of happiness, because we assume that it is through them that we will be happy. On the other hand, no one chooses happiness for the sake of honor, pleasure, and the like, nor as a means to anything at all.

We arrive at the same conclusion if we approach the question from the standpoint of self-sufficiency. For the final and perfect good seems to be self-sufficient. However, we define something as self-sufficient not by reference to the "self" alone. We do not mean a man who lives his life in isolation, but a man who also lives with parents, children, a wife, and friends and fellow citizens generally, since man is by nature a social and political being. But some limit must be set to these relationships; for if they are extended to include ancestors, descendants, and friends of friends, they will go on to infinity. . . . For the present we define as "self-sufficient" that which taken by itself makes life something desirable and deficient in nothing. It is happiness, in our opinion, which fits this description. Moreover, happiness is of all things the one most desirable, and it is not counted as one good thing among many others. But if it were counted as one among many others, it is obvious that the addition of even the least of the goods would make it more desirable; for the addition would produce an extra amount of good, and the greater amount of good is always more desirable than the lesser. We see then that happiness is something final and self-sufficient and the end of our actions.

To call happiness the highest good is perhaps a little trite, and a clearer account of what it is, is still required. Perhaps this is best done by first ascertaining the proper function of man. For just as the goodness and performance of a flute player, a sculptor, or any kind of expert, and generally of anyone who fulfills some function or performs some action, are thought to reside in his proper function, so the goodness and performance of man would seem to reside in whatever is his proper function. Is it then possible that while a carpenter and a shoemaker have their own proper functions and spheres of action, man as man has none, but was left by nature a good-for-nothing without a function? Should we not assume that just as the eye, the hand, the foot, and in general each part of the body clearly has its own proper function, so man too has some function over and above the functions of his parts? What can this function possibly be? Simply living? He shares that even with plants, but we are now looking for something peculiar to man. Accordingly, the life of nutrition and growth must be excluded. Next in line there is a life of sense perception. But this, too, man has in common with the horse, the ox, and every animal. There remains then an active life of the rational element. The rational element has two parts: one is rational in that it obeys the rule of reason, the other in that it possesses and conceives rational rules. Since the expression "life of the rational element" also can be used in two senses, we must make it clear that we mean a life determined by the activity, as opposed to the mere possession, of the rational element. For the activity, it seems, has a greater claim to be the function of man.

The proper function of man, then, consists in an activity of the soul in conformity with a rational principle or, at least, not

without it. In speaking of the proper function of a given individual we mean that it is the same in kind as the function
of an individual who sets high standards for himself: the proper
function of a harpist, for example, is the same as the function
of a harpist who has set high standards for himself. The same
applies to any and every group of individuals: the full attainment
of excellence must be added to the mere function. In other words,
the function of the harpist is to play the harp; the function of
the harpist who has high standards is to play it well. On these
assumptions, if we take the proper function of man to be a
certain kind of life, and if this kind of life is an activity of the
soul and consists in actions performed in conjunction with the
rational element, and if a man of high standards is he who
performs these actions well and properly, and if a function is
well performed when it is performed in accordance with the
excellence appropriate to it; we reach the conclusion that the
good of man is an activity of the soul in conformity with excellence or virtue, and if there are several virtues, in conformity
with the best and most complete.

But we must add "in a complete life." For one swallow does
not make a spring, nor does one sunny day; similarly, one day
or a short time does not make a man blessed and happy. . . .

Popular views about happiness confirm our position

We must examine the fundamental principle with which we
are concerned [happiness], not only on the basis of the logical
conclusion we have reached and on the basis of the elements
which make up its definition, but also on the basis of the views
commonly expressed about it. For in a true statement, all the
facts are in harmony; in a false statement, truth soon introduces
a discordant note.

Good things are commonly divided into three classes: (1)
external goods, (2) goods of the soul, and (3) goods of the body.

Of these, we call the goods pertaining to the soul goods in the highest and fullest sense. But in speaking of "soul," we refer to our soul's actions and activities. Thus, our definition tallies with this opinion which has been current for a long time and to which philosophers subscribe. We are also right in defining the end as consisting of actions and activities; for in this way the end is included among the goods of the soul and not among external goods.

Also the view that a happy man lives well and fares well fits in with our definition: for we have all but defined happiness as a kind of good life and well-being.

Moreover, the characteristics which one looks for in happiness are all included in our definition. For some people think that happiness is virtue, others that it is practical wisdom, others that it is some kind of theoretical wisdom; others again believe it to be all or some of these accompanied by, or not devoid of, pleasure; and some people also include external prosperity in its definition. Some of these views are expressed by many people and have come down from antiquity, some by a few men of high prestige, and it is not reasonable to assume that both groups are altogether wrong; the presumption is rather that they are right in at least one or even in most respects.

Now, in our definition we are in agreement with those who describe happiness as virtue or as some particular virtue, for our term "activity in conformity with virtue" implies virtue. But it does doubtless make a considerable difference whether we think of the highest good as consisting in the possession or in the practice of virtue, viz., as being a characteristic or an activity. For a characteristic may exist without producing any good result, as for example, in a man who is asleep or incapacitated in some other respect. An activity, on the other hand, must produce a result: [an active person] will necessarily act and act well. Just as the crown at the Olympic Games is not awarded to the most beautiful and the strongest but to the participants in the contests—for it is among them that the victors are found—so the good and noble things in life are won by those who act rightly.

The life of men active in this sense is also pleasant in itself. For the sensation of pleasure belongs to the soul, and each man derives pleasure from what he is said to love: a lover of horses from horses, a lover of the theater from plays, and in the same way a lover of justice from just acts, and a lover of virtue in general from virtuous acts. In most men, pleasant acts conflict with one another because they are not pleasant by nature, but men who love what is noble derive pleasure from what is naturally pleasant. Actions which conform to virtue are naturally pleasant, and, as a result, such actions are not only pleasant for those who love the noble but also pleasant in themselves. The life of such men has no further need of pleasure as an added attraction, but it contains pleasure within itself. We may even go so far as to state that the man who does not enjoy performing noble actions is not a good man at all. Nobody would call a man just who does not enjoy acting justly, nor generous who does not enjoy generous actions, and so on. If this is true, actions performed in conformity with virtue are in themselves pleasant.

Of course it goes without saying that such actions are good as well as noble, and they are both in the highest degree, if the man of high moral standards displays any right judgment about them at all; and his judgment corresponds to our description. So we see that happiness is at once the best, noblest, and most pleasant thing, and these qualities are not separate, as the inscription at Delos makes out:

> The most just is most noble, but health is the best,
> and to win what one loves is pleasantest.

For the best activities encompass all these attributes, and it is in these, or in the best one of them, that we maintain happiness consists.

Still, happiness, as we have said, needs external goods as well. For it is impossible or at least not easy to perform noble actions if one lacks the wherewithal. Many actions can only be performed with the help of instruments, as it were: friends, wealth, and political power. And there are some external goods the absence of which spoils supreme happiness, e.g., good birth, good children, and beauty: for a man who is very ugly in appearance or ill-born or who lives all by himself and has no children cannot be classified as altogether happy; even less happy perhaps is a man whose children and friends are worthless, or one who has lost good children and friends through death. Thus, as we have said, happiness also requires well-being of this kind, and that is the reason why some classify good fortune with happiness, while others link it to virtue.

How happiness is acquired

This also explains why there is a problem whether happiness is acquired by learning, by discipline, or by some other kind of training, or whether we attain it by reason of some divine dispensation or even by chance. Now, if there is anything at all which comes to men as a gift from the gods, it is reasonable to suppose that happiness above all else is god-given; and of all things human it is the most likely to be god-given, inasmuch as it is the best. But although this subject is perhaps more appropriate to a different field of study, it is clear that happiness is one of the most divine things, even if it is not god-sent but attained through virtue and some kind of learning or training. For the prize and end of excellence and virtue is the best thing of all, and it is something divine and blessed. Moreover, if happiness depends on excellence, it will be shared by many people; for study and effort will make it accessible to anyone whose capacity for virtue is unimpaired. And if it is better that happiness is acquired in this way rather than by chance, it is reasonable to assume that this is the way in which it is acquired. For, in the realm of nature, things are naturally arranged in the best way possible—and the same is also true of the products of art and of any kind of causation, especially the highest. To leave the greatest and noblest of things to chance would hardly be right.

A solution of this question is also suggested by our earlier definition, according to which the good of man, happiness, is some kind of activity of the soul in conformity with virtue. All the other goods are either necessary prerequisites for happiness, or are by nature co-workers with it and useful instruments for attaining it. Our results also tally with what we said at the outset: for we stated that the end of politics is the best of ends; and the main concern of politics is to engender a certain character in the citizens and to make them good and disposed to perform noble actions.

We are right, then, when we call neither a horse nor an ox nor any other animal happy, for none of them is capable of participating in an activity of this kind. For the same reason, a child is not happy, either; for, because of his age, he cannot yet perform such actions. When we do call a child happy, we do so by reason of the hopes we have for his future. Happiness, as we have said, requires completeness in virtue as well as a complete lifetime. Many changes and all kinds of contingencies befall a man in the course of his life, and it is possible that the most prosperous man will encounter great misfortune in his old age, as the Trojan legends tell about Priam. When a man has met a fate such as his and has come to a wretched end, no one calls him happy.

Can a man be called "happy" during his lifetime?

Must we, then, apply the term "happy" to no man at all as long as he is alive? Must we, as Solon would have us do, wait to see his end? And, on this assumption, is it also true that a man is actually happy after he is dead? Is this not simply absurd, especially for us who define happiness as a kind of activity? Suppose we do not call a dead man happy, and interpret Solon's words to mean that only when a man is dead can we safely say that he has been happy, since he is now beyond the reach of evil and misfortune—this view, too, is open to objection. For it seems that to some extent good and evil really exist for a dead man, just as they may exist for a man who lives without being conscious of them, for example, honors and disgraces, and generally the successes and failures of his children and descendants. This presents a further problem. A man who has lived happily to his old age and has died as happily as he lived may have many vicissitudes befall his descendants: some of them may be good and may be granted the kind of life which they deserve, and others may not. It is, further, obvious that the descendants may conceivably be removed from their ancestors by various degrees. Under such circumstances, it would be odd if the dead man would share in the vicissitudes of his descendants and be happy at one time and wretched at another. But it would also be odd if the fortunes of their descendants did not affect the ancestors at all, not even for a short time.

But we must return to the problem raised earlier, for through it our present problem perhaps may be solved. If one must look to the end and praise a man not as being happy but as having been happy in the past, is it not paradoxical that at a time when a man actually is happy this attribute, though true, cannot be applied to him? We are unwilling to call the living happy because changes may befall them and because we believe that happiness has permanence and is not amenable to changes under any circumstances, whereas fortunes revolve many times in one person's lifetime. For obviously, if we are to keep pace with a man's fortune, we shall frequently have to call the same man happy at one time and wretched at another and demonstrate that the happy man is a kind of chameleon, and that the foundations [of his life] are unsure. Or is it quite wrong to make our judgment depend on fortune? Yes, it is wrong, for fortune does not determine whether we fare well or ill, but is, as we said, merely an accessory to human life; activities in conformity with virtue constitute happiness, and the opposite activities constitute its opposite.

The question which we have just discussed further confirms our definition. For no function of man possesses as much stability as do activities in conformity with virtue: these seem to be even more durable than scientific knowledge. And the higher the virtuous activities, the more durable they are, because men who are supremely happy spend their lives in these activities most intensely and most continuously, and this seems to be the reason why such activities cannot be forgotten.

The happy man will have the attribute of permanence which we are discussing, and he will remain happy throughout his life. For he will always or to the highest degree both do and contemplate what is in conformity with virtue; he will bear the vicissitudes of fortune most nobly and with perfect decorum under all circumstances, inasmuch as he is truly good and "foursquare beyond reproach."

But fortune brings many things to pass, some great and some small. Minor instances of good and likewise of bad luck obviously do not decisively tip the scales of life, but a number of major successes will make life more perfectly happy; for, in the first place, by their very nature they help to make life attractive, and secondly, they afford the opportunity for noble and good actions. On the other hand, frequent reverses can crush and mar supreme happiness in that they inflict pain and thwart many activities. Still, nobility shines through even in such circumstances, when a man bears many great misfortunes with good grace not because he is insensitive to pain but because he is noble and high-minded.

If, as we said, the activities determine a man's life, no supremely happy man can ever become miserable, for he will never do what is hateful and base. For in our opinion, the man who is truly good and wise will bear with dignity whatever fortune may bring, and will always act as nobly as circumstances permit, just as a good general makes the most strategic use of the troops at his disposal, and a good shoemaker makes the best shoe he can from the leather available, and so on with experts in all other fields. If this is true, a happy man will never become miserable; but even so, supreme happiness will not be his if a fate such as Priam's befalls him. And yet, he will not be fickle and changeable; he will not be dislodged from his happiness easily by any misfortune that comes along, but only by great and numerous disasters such as will make it impossible for him to become happy again in a short time; if he recovers his happiness at all, it will be only after a long period of time, in which he has won great distinctions.

Is there anything to prevent us, then, from defining the happy man as one whose activities are an expression of complete virtue, and who is sufficiently equipped with external goods, not simply at a given moment but to the end of his life? Or should we add that he must die as well as live in the manner which we have defined? For we cannot foresee the future, and happiness, we maintain, is an end which is absolutely final and complete in every respect. If this be granted, we shall define as "supremely happy" those living men who fulfill and continue to fulfill these requirements, but blissful only as human beings. So much for this question.

Do the fortunes of the living affect the dead?

That the fortunes of his descendants and of all those near and dear to him do not affect the happiness of a dead man at all, seems too unfeeling a view and contrary to the prevailing opinions. Many and different in kind are the accidents that can befall us, and some hit home more closely than others. It would, therefore, seem to be a long and endless task to make detailed distinctions, and perhaps a general outline will be sufficient. Just as one's own misfortunes are sometimes momentous and decisive for one's life and sometimes seem comparatively less important, so the misfortunes of our various friends affect us to varying degrees. In each case it makes a considerable difference whether those who are affected by an event are living or dead; much more so than it matters in a tragedy whether the crimes and horrors have been perpetrated before the opening of the play or are part of the plot. This difference, too, must be taken into account and perhaps still more the problem whether the dead participate in any good or evil. These considerations suggest that even if any good or evil reaches them at all, it must be something weak and negligible (either intrinsically or in relation to them), or at least something too small and insignificant to make the unhappy happy or to deprive the happy of their bliss. The good as well as the bad fortunes of their friends seem, then, to have some effect upon the dead, but the nature and magnitude of the effect is such as not to make the happy unhappy or to produce any similar changes. . . .

The psychological foundations of the virtues

Since happiness is a certain activity of the soul in conformity with perfect virtue, we must now examine what virtue or excellence is. For such an inquiry will perhaps better enable us to discover the nature of happiness. Moreover, the man who is truly concerned about politics seems to devote special attention to excellence, since it is his aim to make the citizens good and law-abiding. We have an example of this in the lawgivers of Crete and Sparta and in other great legislators. If an examination of virtue is part of politics, this question clearly fits into the pattern of our original plan.

There can be no doubt that the virtue which we have to study is human virtue. For the good which we have been seeking is a human good and the happiness a human happiness. By human virtue we do not mean the excellence of the body, but that of the soul, and we define happiness as an activity of the soul. If this is true, the student of politics must obviously have some knowledge of the workings of the soul, just as the man who is to heal eyes must know something about the whole body. In fact, knowledge is all the more important for the former, inasmuch as politics is better and more valuable than medicine, and cultivated physicians devote much time and trouble to gain knowledge about the body. . . .

The soul consists of two elements, one irrational and one rational. Whether these two elements are separate, like the parts of the body or any other divisible thing, or whether they are only logically separable though in reality indivisible, as convex and concave are in the circumference of a circle, is irrelevant for our present purposes.

Of the irrational element, again, one part seems to be common to all living things and vegetative in nature: I mean that part which is responsible for nurture and growth. We must assume that some such capacity of the soul exists in everything that takes nourishment, in the embryonic stage as well as when the organism is fully developed; for this makes more sense than to assume the existence of some different capacity at the latter stage. The excellence of this part of the soul is, therefore, shown to be common to all living things and is not exclusively human. This very part and this capacity seem to be most active in sleep. For in sleep the difference between a good man and a bad is least apparent—whence the saying that for half their lives the happy are no better off than the wretched. This is just what we would expect, for sleep is an inactivity of the soul in that it ceases to do things which cause it to be called good or bad. However, to a small extent some bodily movements do penetrate to the soul in sleep, and in this sense the dreams of honest men are better than those of average people. But enough of this subject: we may pass by the nutritive part, since it has no natural share in human excellence or virtue.

In addition to this, there seems to be another integral element of the soul which, though irrational, still does partake of reason in some way. In morally strong and morally weak men we praise the reason that guides them and the rational element of the soul, because it exhorts them to follow the right path and to do what is best. Yet we see in them also another natural strain different from the rational, which fights and resists the guidance of reason. The soul behaves in precisely the same manner as do the paralyzed limbs of the body. When we intend to move the limbs to the right, they turn to the left, and similarly, the impulses of morally weak persons turn in the direction opposite to that in which reason leads them. However, while the aberration of the body is visible, that of the soul is not. But perhaps we must accept it as a fact, nevertheless, that there is something in the soul besides the rational element, which opposes and reacts against it. In what way the two are distinct need not concern us here. But, as we have stated, it too seems to partake of reason; at any rate, in a morally strong man it accepts the leadership of reason, and is perhaps more obedient still in a self-controlled and courageous man, since in him everything is in harmony with the voice of reason.

Thus we see that the irrational element of the soul has two parts: the one is vegetative and has no share in reason at all, the other is the seat of the appetites and of desire in general and partakes of reason insofar as it complies with reason and accepts its leadership; it possesses reason in the sense that we say it is "reasonable" to accept the advice of a father and of friends, not in the sense that we have a "rational" understanding of mathematical propositions. That the irrational element can be persuaded by the rational is shown by the fact that admonition and all manner of rebuke and exhortation are possible. If it is correct to say that the appetitive part, too, has reason, it follows that the rational element of the soul has two subdivisions: the one possesses reason in the strict sense, contained within itself, and the other possesses reason in the sense that it listens to reason as one would listen to a father.

Virtue, too, is differentiated in line with this division of the soul. We call some virtues "intellectual" and others "moral": theoretical wisdom, understanding, and practical wisdom are intellectual virtues, generosity and self-control moral virtues. In speaking of a man's character, we do not describe him as wise or understanding, but as gentle or self-controlled; but we praise the wise man, too, for his characteristic, and praiseworthy characteristics are what we call virtues.

BOOK X

Happiness, intelligence, and the contemplative life

. . . Now, if happiness is activity in conformity with virtue, it is to be expected that is should conform with the highest virtue, and that is the virtue of the best part of us. Whether this is intelligence or something else which, it is thought, by its very nature rules and guides us and which gives us our notions of what is noble and divine; whether it is itself divine or the most divine thing in us; it is the activity of this part [when operating] in conformity with the excellence or virtue proper to it that will be complete happiness. That it is an activity concerned with theoretical knowledge or contemplation has already been stated.

This would seem to be consistent with our earlier statements as well as the truth. For this activity is not only the highest—for intelligence is the highest possession we have in us, and the objects which are the concern of intelligence are the highest objects of knowledge—but also the most continuous: we are able to study continuously more easily than to perform any kind of action. Furthermore, we think of pleasure as a necessary ingredient in happiness. Now everyone agrees that of all the activities that conform with virtue activity in conformity with theoretical wisdom is the most pleasant. At any rate, it seems that [the pursuit of wisdom or] philosophy holds pleasures marvelous in purity and certainty, and it is not surprising that time spent in knowledge is more pleasant than time spent in research. Moreover, what is usually called "self-sufficiency" will be found in the highest degree in the activity which is concerned with theoretical knowledge. Like a just man and any other virtuous man, a wise man requires the necessities of life; once these have been adequately provided, a just man still needs people toward whom and in company with whom to act justly, and the same is true of a self-controlled man, a courageous man, and all the rest. But a wise man is able to study even by himself, and the wiser he is the more is he able to do it. Perhaps he could do it better if he had colleagues to work with him, but he still is the most self-sufficient of all. Again, study seems to be the only activity which is loved for its own sake. For while we derive a greater or a smaller advantage from practical pursuits beyond the action itself, from study we derive nothing beyond the activity of studying. Also, we regard happiness as depending on leisure; for our purpose in being busy is to have leisure, and we wage war in order to have peace. Now, the practical virtues are activated in political and military pursuits, but the actions involved in these pursuits seem to be unleisurely. This is completely true of military pursuits, since no one chooses to wage war or foments war for the sake of war; he would have to be utterly bloodthirsty if he were to make enemies of his friends simply in order to have battle and slaughter. But the activity of the statesman, too, has no leisure. It attempts to gain advantages beyond political action, advantages such as political power, prestige, or at least happiness for the statesman himself and his fellow citizens, and that is something other than political activity: after all, the very fact that we investigate politics shows that it is not the same [as happiness]. Therefore, if we take as established (1) that political and military actions surpass all other actions that conform with virtue in nobility and grandeur; (2) that they are unleisurely, aim at an end, and are not chosen for their own sake; (3) that the activity of our intelligence, inasmuch as it is an activity concerned with theoretical knowledge, is thought to be of greater value than the others, aims at no end beyond itself, and has a pleasure proper to itself—and pleasure increases activity; and (4) that the qualities of this activity evidently are self-sufficiency, leisure, as much freedom from fatigue as a human being can have, and whatever else falls to the lot of a supremely happy man; it follows that the activity of our intelligence constitutes the complete happiness of man, provided that it encompasses a complete span of life; for nothing connected with happiness must be incomplete.

However, such a life would be more than human. A man who would live it would do so not insofar as he is human, but because there is a divine element within him. This divine element is as far above our composite nature[3] as its activity is above the active exercise of the other [i.e., practical] kind of virtue. So if it is true that intelligence is divine in comparison with man, then a life guided by intelligence is divine in comparison with human life. We must not follow those who advise us to have human thoughts, since we are [only] men, and mortal thoughts, as mortals should; on the contrary, we should try to become immortal as far as that is possible and do our utmost to live in accordance with what is highest in us. For though this is a small portion [of our nature], it far surpasses everything

---

[3] [Man, consisting of soul and body, i.e., of form and matter, is a composite being, whereas the divine, being all intelligence, is not.—TRANS.]

else in power and value. One might even regard it as each man's true self, since it is the controlling and better part. It would, therefore, be strange if a man chose not to live his own life but someone else's.

Moreover, . . . what is by nature proper to each thing will be at once the best and the most pleasant for it. In other words, a life guided by intelligence is the best and most pleasant for man, inasmuch as intelligence, above all else, is man. Consequently, this kind of life is the happiest.

PLATO was born in Athens, Greece in about 428 B.C. One family ancestor was said to be the Greek god Poseidon, and another was Solon, the distinguished Athenian lawmaker and reformer. Plato's parents were prominent in Athenian affairs; through them he became acquainted with the philosopher Socrates. As a young man, Plato considered pursuing a political career. But the violence and corruption he observed in Athenian politics—he witnessed the destruction of the Athenian empire in the Peloponnesian War and the civil disruption that followed this—dissuaded him from his political ambitions. His association with the followers of Socrates led Plato to found the Academy, a school dedicated to philosophical and scientific research in Athens, in 387 B.C. The Academy survived for 900 years. Plato taught and officiated there—while also writing his 36 dialogues —until his death in approximately 348 B.C.

From *Socrates and Legal Obligation,* translated by R. E. Allen. Publisher: The University of Minnesota Press, 1980. Pages 37–62.

The Apology

To what degree, Gentlemen of Athens, you have been affected by my accusers, I do not know. I, at any rate, was almost led to forget who I am — so convincingly did they speak. Yet hardly anything they have said is true. Among their many falsehoods, I was especially surprised by one; they said you must be on guard lest I deceive you, since I am a clever speaker. To have no shame at being directly refuted by facts when I show myself in no way clever with words — that, I think, is the very height of shamelessness. Unless, of course, they call a man a clever speaker if he speaks the truth. If that is what they mean, why, I would even admit to being an orator — though not after *their* fashion.

These men, I claim, have said little or nothing true. But from me, Gentlemen, you will hear the whole truth. It will not be prettily tricked out in elegant speeches like theirs, words and phrases all nicely arranged. To the contrary: you will hear me speak naturally in the words which happen to occur to me. For I believe what I say to be just, and let no one of you expect otherwise. Besides, it would hardly be appropriate in a man of my age, Gentlemen, to come before you making up speeches like a boy. So I must specifically ask one thing of you, Gentlemen. If you hear me make my defense in the same words I customarily use at the tables in the Agora, and other places where many of you have heard me, please do not be surprised or make a disturbance because of it. For things stand thus: I am now come into court for the first time; I am seventy years old; and I am an utter stranger to this place. If I were a foreigner, you would unquestionably make allowances if I spoke in the dialect and manner in which I was raised. In just the same way, I specifically ask you now, and justly so, I think, to pay no attention to my manner of speech—it may perhaps be poor, but then perhaps an improvement—and look strictly to this one thing, whether or not I speak justly. For that is the virtue of a judge, and the virtue of an orator is to speak the truth.

First of all, Gentlemen, it is right for me to defend myself against the first false accusations lodged against me, and my first accusers; and next, against later accusations and later accusers. For the fact is that many accusers have risen before you against me; at this point they have been making accusations for many years, and they have told no truth. Yet I fear them more than I fear Anytus and those around him—though they too are dangerous. Still, the others are more dangerous. They took hold of most of you in childhood, persuading you of the truth of accusations which were in fact quite false: "There is a certain Socrates . . . Wise man . . . Thinker on things in the Heavens . . . Inquirer into all things beneath Earth . . . Making the weaker argument stronger. . . . " Those men, Gentlemen of Athens, the men who spread that report, are my dangerous accusers; for their hearers believe that those who inquire into such things acknowledge no gods.

Again, there have been many such accusers, and they have now been at work for a long time; they spoke to you at a time when you were especially trusting—some of you children, some only a little older—and they lodged their accusations quite by default, no one appearing in defense. But the most absurd thing is that one cannot even know or tell their names—unless perhaps in the case of a comic poet. But those who use malicious slander to persuade you, and those who, themselves persuaded, persuade others—all are most difficult to deal with. For it is impossible to bring any one of them forward as a witness and cross-examine him. I must rather, as it were, fight with shadows in making my defense, and question where no one answers.

Please grant, then, as I say, that two sets of accusers have risen against me: those who now lodge their accusations, and those who lodged accusations long since. And please accept the fact that I must defend myself against the latter first. For in fact, you heard their accusations earlier, and with far greater effect than those which came later.

Very well then. A defense is to be made, Gentlemen of Athens. I am to attempt to remove from you in this short time that prejudice which you have been long in acquiring. I might wish that this should come to pass, if it were in some way better for you and for me, wish that I might succeed in my defense. But I think the thing difficult, and its nature hardly escapes me. Still, let that go as pleases the God; the law must be obeyed, and a defense conducted.

Let us then take up from the beginning the charges which have given rise to the prejudice — the charges on which Meletus in fact relied in lodging his indictment. Very well, what do those who slander me say? It is necessary to read, as it were, their sworn indictment: "Socrates is guilty of needless curiosity and meddling interference, inquiring into things beneath Earth and in the Sky, making the weaker argument stronger, and teaching others the same." The charge is something like that. Indeed, you have seen it for yourselves in a comedy by Aristophanes — a certain Socrates being carried around on the stage, talking about walking on air and babbling a great deal of other nonsense, of which I understand neither much nor little. Mark you, I do not mean to disparage such knowledge, if anyone in fact has it — let me not be brought to trial by Meletus on such a charge as that! But Gentlemen, I have no share of it. Once again, I offer the majority of you as witnesses, and ask those of you who have heard me in conversation — there are many among you —

inform each other, please, whether any of you ever heard anything of the sort. From that you will recognize the nature of
the other things the multitude says about me.

The fact is that there is nothing in these accusations. And if
you have heard from anyone that I undertake to educate men,
and make money doing it, that is false too. Once again, I think
it would be a fine thing to be able to educate men, as Gorgias
of Leontini does, or Prodicus of Ceos, or Hippias of Elis. For
each of them, Gentlemen, can enter any given city and convince
the youth—who might freely associate with any of their fellow
citizens they please—to drop those associations and associate
with them, to pay money for it, and give thanks in the bargain.
As a matter of fact, there is a man here right now, a Parian,
and a wise one, who as I learn has just come to town. For I
happened to meet a person who has spent more money on
Sophists than everyone else put together, Callias, son of Hip-
ponicus. So I asked him—for he has two sons—"Callias," I
said, "if your two sons were colts or calves, we could get an
overseer for them and hire him, and his business would be to
make them excellent in their appropriate virtue. He would be
either a horse-trainer or a farmer. But as it is, since the two of
them are men, whom do you intend to get as an overseer? Who
has knowledge of that virtue which belongs to a man and a
citizen? Since you have sons, I'm sure you have considered this.
Is there such a person," I said, "or not?"

"To be sure," he said.

"Who is he?" I said. "Where is he from, and how much
does he charge to teach?"

"Evenus, Socrates," he said. "A Parian. Five minae."

And I count Evenus fortunate indeed, if he really possesses
that art, and teaches it so modestly. For my own part, at any
rate, I would be puffed up with vanity and pride if I had such
knowledge. But I do not, Gentlemen.

Perhaps one of you will ask, "But Socrates, what is this all about? Whence have these slanders against you arisen? You must surely have been busying yourself with something out of the ordinary; so grave a report and rumor would not have arisen had you not been doing something rather different from most folk. Tell us what it is, so that we may not take action in your case unadvisedly." That, I think, is a fair request, and I shall try to indicate what it is that has given me the name I have. Hear me, then. Perhaps some of you will think I joke; be well assured that I shall be telling the whole truth.

Gentlemen of Athens, I got this name through nothing but a kind of wisdom. What kind? The kind which is perhaps peculiarly human, for it may be I am really wise in that. And perhaps the men I just mentioned are wise with a wisdom greater than human—either that, or I cannot say what. In any case, I have no knowledge of it, and whoever says I do is lying and speaks to my slander.

Please, Gentlemen of Athens. Do not make a disturbance, even if I seem to you to boast. For it will not be my own words I utter; I shall refer you to the speaker, as one worthy of credit. For as witness to you of my wisdom—whether it is wisdom of a kind, and what kind of wisdom it is—I shall call the God at Delphi.

You surely knew Chaerephon. He was my friend from youth, and a friend of your democratic majority. He went into exile with you, and with you he returned. And you know what kind of a man he was, how eager and impetuous in whatever he rushed into. Well, he once went to Delphi and boldly asked the oracle—as I say, Gentlemen, please do not make a disturbance—he asked whether anyone is wiser than I. Now, the Pythia replied that no one is wiser. And to this his brother here will testify, since Chaerephon is dead.

Why do I mention this? I mention it because I intend to inform you whence the slander against me has arisen. For when

I heard it, I reflected: "What does the God mean? What is the sense of this riddling utterance? I know that I am not wise at all; what then does the God mean by saying I am wisest? Surely he does not speak falsehood; it is not permitted to him." So I puzzled for a long time over what was meant, and then, with great reluctance, I turned to inquire into the matter in some such way as this.

I went to someone with a reputation for wisdom, in the belief that there if anywhere I might test the meaning of the utterance and declare to the oracle that, "This man is wiser than I am, and you said I was wisest." So I examined him—there is no need to mention a name, but it was someone in political life who produced this effect on me in discussion, Gentlemen of Athens—and I concluded that though he seemed wise to many other men, and most especially to himself, he was not. I tried to show him this; and thence I became hated, by him and by many who were present. But I left thinking to myself, "I am wiser than that man. Neither of us probably knows anything worthwhile; but he thinks he does and does not, and I do not and do not think I do. So it seems at any rate that I am wiser in this one small respect: I do not think I know what I do not." I then went to another man who was reputed to be even wiser, and the same thing seemed true again; there too I became hated, by him and by many others.

Nevertheless, I went on, perceiving with grief and fear that I was becoming hated, but still, it seemed necessary to put the God first—so I had to go on, examining what the oracle meant by testing everyone with a reputation for knowledge. And by the Dog, Gentlemen—I must tell you the truth—I swear that I had some such experience as this: it seemed to me that those most highly esteemed for wisdom fell little short of being most deficient, as I carried on inquiry in behalf of the God, and that others reputedly inferior were men of more discernment.

But really, I must display my wanderings to you; they were

like those of a man performing labors—all to the end that I might not leave the oracle untested. From the politicians I went to the poets—tragic, dithyrambic, and the rest—thinking that there I would discover myself manifestly less wise by comparison. So I took up poems over which I thought they had taken special pains, and asked them what they meant, so as also at the same time to learn from them. Now, I am ashamed to tell you the truth, Gentlemen, but still, it must be told. There was hardly anyone present who could not give a better account than they of what they had themselves produced. So presently I came to realize that poets do not make what they make by wisdom, but by a kind of native disposition or divine inspiration, exactly like seers and prophets. For the latter also utter many fine things, but know nothing of the things of which they speak. That is how the poets also appeared to me, while at the same time I realized that because of their poetry they thought themselves the wisest of men in other matters—and were not. Once again, I left thinking myself superior to them in just the way I was to the politicians.

Finally I went to the craftsmen. I was aware that although I knew scarcely anything, I would find that they knew many things, and fine ones. In this I was not mistaken: they knew things that I did not, and in that respect were wiser. But Gentlemen of Athens, it seemed to me that the poets and our capable public craftsmen had exactly the same failing: because they practiced their own arts well, each deemed himself wise in other things, things of great importance. This mistake quite obscured their wisdom. The result was that I asked myself on behalf of the oracle whether I would accept being such as I am, neither wise with their wisdom nor foolish with their folly, or whether I would accept wisdom and folly together and become such as they are. I answered, both for myself and the oracle, that it was better to be as I am.

From this examination, Gentlemen of Athens, much enmity

has risen against me, of a sort most harsh and heavy to endure, so that many slanders have arisen, and the name is put abroad that I am "wise." For on each occasion those present think I am wise in the things in which I test others. But very likely, Gentlemen, it is really the God who is wise, and by his oracle he means to say that, "Human nature is a thing of little worth, or none." It appears that he does not mean this fellow Socrates, but uses my name to offer an example, as if he were saying that, "He among you, Gentlemen, is wisest who, like Socrates, realizes that he is truly worth nothing in respect to wisdom." That is why I still go about even now on behalf of the God, searching and inquiring among both citizens and strangers, should I think some one of them is wise; and when it seems he is not, I help the God and prove it. Due to this pursuit, I have no leisure worth mentioning either for the affairs of the City or for my own estate; I dwell in utter poverty because of my service to God.

Then too the young men follow after me—especially the ones with leisure, namely, the richest. They follow of their own initiative, rejoicing to hear men tested, and often they imitate me and undertake to test others; and next, I think, they find an ungrudging plenty of people who think they have some knowledge but know little or nothing. As a result, those whom they test become angry at me, not at themselves, and say that, "This fellow Socrates is utterly polluted, and corrupts the youth." And when someone asks them what it is he does, what it is he teaches, they cannot say because they do not know; but so as not to seem at a loss, they mutter the kind of thing that lies ready to hand against anyone who pursues wisdom: "Things in the Heavens and beneath the Earth," or, "Not acknowledging gods," or, "Making the weaker argument stronger." The truth, I suppose, they would not wish to state, namely, that it is become quite clear that they pretend to knowledge and know nothing. And because they are concerned for their pride, I think, and zealous, and numerous, and speak vehemently and persuasively about me, they have long filled your ears with zealous slander. It was on the strength of this that Meletus attacked me, along with Anytus and Lycon — Meletus angered on behalf of the poets, Anytus on behalf of the public craftsmen and the politicians, Lycon on behalf of the orators. So the result is, as I said to begin with, that I should be most surprised were I able to remove from you in this short time a slander which has grown so great. There, Gentlemen of Athens, you have the truth, and I have concealed or misrepresented nothing in speaking it, great or small. Yet I know quite well that it is just for this that I have become hated — which is in fact an indication of the truth of what I say, and that this is the basis of the slander and charges against me. Whether you inquire into it now or hereafter you will find it to be so.

Against the charges lodged by my first accusers, let this defense suffice. But for Meletus — the good man who loves his City, so he says — and for my later accusers, I shall attempt a further defense. Once more then, as before a different set of accusers, let us take up their sworn indictment. It runs something like this: it says that Socrates is guilty of corrupting the youth, and of not acknowledging the gods the City acknowledges, but other new divinities. Such is the charge. Let us examine its particulars.

It claims I am guilty of corrupting the youth. But I claim, Gentlemen of Athens, that it is Meletus who is guilty — guilty of jesting in earnest, guilty of lightly bringing men to trial, guilty of pretending a zealous concern for things he never cared about at all. I shall try to show you that this is true.

Come here, Meletus. Now tell me. Do you count it of great importance that the young should be as good as possible?

"I do."

Then come and tell the jurors this: who improves them? Clearly you know, since it is a matter of concern to you. Having discovered, so you say, that I am the man who is corrupting them, you bring me before these judges to accuse me. But now come and say who makes them better. Inform the judges who he is.

You see, Meletus. You are silent. You cannot say. And yet, does this not seem shameful to you, and a sufficient indication of what I say, namely, that you never cared at all? Tell us, my friend. Who improves them?

"The laws."

But I did not ask you that, dear friend. I asked you what man improves them—whoever it is who in the first place knows just that very thing, the laws.

"These men, Socrates. The judges."

Really Meletus? These men here are able to educate the youth and improve them?

"Especially they."

All of them? Or only some?

"All."

By Hera, you bring good news. An ungrudging plenty of benefactors! But what about the audience here. Do they improve them, or not?

"They too."

And members of the Council?

"The Councillors too."

Well then Meletus, do the members of the Assembly, the Ecclesiasts, corrupt the young? Or do they all improve them too?

"They too."

So it seems that every Athenian makes them excellent except me, and I alone corrupt them. Is that what you are saying?

"That is exactly what I am saying."

You condemn me to great misfortune. But tell me, do you think it is so with horses? Do all men improve them, while some one man corrupts them? Or quite to the contrary, is it some one man or a very few, namely horse-trainers, who are able to improve them, while the majority of people, if they handle horses and use them, corrupt them? Is that not true, Meletus, both of horses and all other animals? Of course it is, whether you and Anytus affirm or deny it. It would be good fortune indeed for the youth if only one man corrupted them and the rest benefited. But the fact is, Meletus, that you sufficiently show that you never gave thought to the youth; you clearly indicate your own lack of concern, indicate that you never cared at all about the matters in which you bring action against me.

But again, dear Meletus, tell us this: is it better to dwell among fellow citizens who are good, or wicked? Do answer, dear friend; surely I ask nothing hard. Do not wicked men do evil things to those around them, and good men good things?

"Of course."

Now, is there anyone who wishes to be harmed rather than benefited by those with whom he associates? Answer me, dear friend, for the law requires you to answer. Is there anyone who wishes to be harmed?

"Of course not."

Very well then, are you bringing action against me here because I corrupt the youth intentionally, or unintentionally?

"Intentionally, I say."

How can that be, Meletus? Are you at your age so much wiser than I at mine that you recognize that evil men always do evil things to those around them, and good men do good, while I have reached such a pitch of folly that I am unaware that if I do some evil to those with whom I associate, I shall very likely receive some evil at their hands, with the result that I do such great evil intentionally, as you claim? I do not believe you, Meletus, and I do not think anyone else does either. On the contrary: either I do not corrupt the youth, or if I do, I do so unintentionally. In either case, you lie. And if I corrupt them unintentionally, it is not the law to bring action here for that sort of mistake, but rather to instruct and admonish in private; for clearly, if I once learn, I shall stop what I unintentionally do. You, however, were unwilling to associate with me and teach me; instead, you brought action here, where it is law to bring those in need of punishment rather than instruction.

Gentlemen of Athens, what I said is surely now clear: Meletus was never concerned about these matters, much or little. Still, Meletus, tell us this: how do you say I corrupt the youth? Or is it clear from your indictment that I teach them not to acknowledge gods the City acknowledges, but other new divinities? Is this what you mean by saying I corrupt by teaching?

"Certainly. That is exactly what I mean."

Then in the name of these same gods we are now discussing, Meletus, please speak a little more plainly still, both for me and for these gentlemen here. Do you mean that I teach the youth to acknowledge that there are gods, and thus do not myself wholly deny gods, and am not in that respect guilty—though the gods are not those the City acknowledges, but different ones? Or are you claiming that I do not myself acknowledge any gods at all, and teach this to others?

"I mean that. You acknowledge no gods at all."

Ah, my dear Meletus, why do you say such things? Do I not at least acknowledge Sun and Moon as gods, as other men do?

"No, no, Gentlemen and Judges, not when he says the Sun is a stone and the Moon earth."

My dear Meletus! Do you think it is Anaxagoras you are accusing? Do you so despise these judges here and think them so unlettered that they do not know it is the books of Anaxagoras of Clazomenae which teem with such statements? Are young men to learn these things specifically from me, when they can buy them sometimes in the Orchestra for a drachma, if the price is high, and laugh at Socrates if he pretends they are his own—especially since they are so absurd? Well, dear friend, is that what you think? I acknowledge no gods at all?

"No, none whatever."

You cannot be believed, Meletus—even, I think, by yourself. Gentlemen of Athens, I think this man who stands here before you is insolent and unchastened, and has brought this suit precisely out of insolence and unchastened youth. He seems to be conducting a test by propounding a riddle: "Will Socrates, the wise man, realize how neatly I contradict myself, or will I deceive him and the rest of the audience?" For certainly it seems clear that he is contradicting himself in his indictment. It is as though he were saying, "Socrates is guilty of not acknowledging gods, and acknowledges gods." Yet surely this is to jest.

Please join me, Gentlemen, in examining why it appears to me that this is what he is saying. And you answer us, Meletus. The rest of you will please remember what I asked you at the beginning, and make no disturbance if I fashion arguments in my accustomed way.

Is there any man, Meletus, who acknowledges that there are things pertaining to men, but does not acknowledge that there are men? Let him answer for himself, Gentlemen—and let him stop interrupting. Is there any man who does not acknowledge that there are horses, but acknowledges things pertaining to horsemanship? Or does not acknowledge that there are flutes, but acknowledges things pertaining to flute playing? There is not, my good friend. If you do not wish to answer, I'll answer for you and for the rest of these people here. But do please answer my next question, at least: Is there any man who acknowledges that there are things pertaining to divinities, but does not acknowledge that there are divinities?

"There is not."

How obliging of you to answer—reluctantly, and under compulsion from these gentlemen here. Now, you say that I acknowledge and teach things pertaining to divinities—whether new or old, still at least I acknowledge them, by your account; indeed, you swore to that in your indictment. But if I acknowledge that there are things pertaining to divinities, must I surely not also acknowledge that there are divinities? Isn't that so? Of course it is—since you do not answer, I count you as agreeing. And divinities, we surely believe, are either gods or children of gods? Correct?

"Of course."

So if I believe in divinities, as you say, and if divinities are a kind of god, there is the jesting riddle I attributed to you: you are saying that I do not believe in gods, and again that I do believe in gods because I believe in divinities. On the other hand, if divinities are children of gods, some born illegitimately of nymphs, or others of whom this is also told, who could possibly believe that there are children of gods, but not gods? It would be as absurd as believing that there are children of horses and asses, namely, mules, without believing there are horses and asses. Meletus, you could not have brought this indictment except in an attempt to test us—or because you were at a loss for any true basis of prosecution. But as to how you are to convince anyone of even the slightest intelligence that one and the same man can believe that there are things pertaining to divinities and gods, and yet believe that there are neither divinities nor heroes—there is no way.

Gentlemen of Athens, I do not think further defense is needed to show that, by the very terms of Meletus' indictment, I am not guilty; this, surely, is sufficient. But as I said before, a great deal of enmity has risen against me among many people, and you may rest assured this is true. And that is what will convict me, if I am convicted—not Meletus, not Anytus, but the grudging slander of the multitude. It has convicted many another good and decent man; I think it will convict me; nor is there reason to fear that with me it will come to a stand.

Perhaps someone may say, "Are you not ashamed, Socrates, at having pursued such a course that you now stand in danger of being put to death?" To him I would make a just reply: You

are wrong, Sir, if you think that a man worth anything at all
should take thought for danger in living or dying. He should
look when he acts to one thing: whether what he does is just
or unjust, the work of a good man or a bad one. By your account,
those demigods and heroes who laid down their lives at Troy
would be of little worth—the rest of them, and the son of Thetis
too; Achilles so much despised danger instead of submitting to
disgrace that when he was intent on killing Hector his goddess
mother told him, as I recall, "My son, if you avenge the slaying
of your comrade Patroclus with the death of Hector, you yourself
shall die; for straightway with Hector is his fate prepared for
you." Achilles heard, and thought little of the death and danger.
He was more afraid to live as a bad man, with friends left
unavenged. "Straightway let me die," he said, "exacting right
from him who did the wrong, that I may not remain here as
a butt of mockery beside the crook-beaked ships, a burden to
the earth." Do you suppose that he gave thought to death and
danger?

Gentlemen of Athens, truly it is so: wherever a man stations
himself in belief that it is best, wherever he is stationed by his
commander, there he must I think remain and run the risks,
giving thought to neither death nor any other thing except
disgrace. I should indeed have wrought a fearful thing, Gentlemen of Athens, if, when the commanders you chose stationed
me at Potidaea and Amphipolis and Delium, I there remained
as others did, and ran the risk of death; but then, when the
God stationed me, as I thought and believed, obliging me to
live in the pursuit of wisdom, examining myself and others—
if then, at that point, through fear of death or any other thing,
I left my post. That would have been dreadful indeed, and then
in truth might I be justly brought to court for not acknowledging
the existence of gods, for willful disobedience to the oracle, for
fearing death, for thinking myself wise when I am not.

For to fear death, Gentlemen, is nothing but to think one is wise when one is not; for it is to think one knows what one does not. No man knows death, nor whether it is not the greatest of all goods; and yet men fear it as though they well knew it to be the worst of evils. Yet how is this not folly most to be reproached, the folly of believing one knows what one does not? I, at least, Gentlemen, am perhaps superior to most men here and just in this, and if I were to claim to be wiser than anyone else it would be in this: that as I have no satisfactory knowledge of things in the Place of the Dead, I do not think I do. I do know that to be guilty of disobedience to a superior, be he god or man, is shameful evil.

So as against evils I know to be evils, I shall never fear or flee from things which for aught I know may be good. Thus, even if you now dismiss me, refusing to do as Anytus bids—Anytus, who said that either I should not have been brought to trial to begin with or, since brought, must be put to death, testifying before you that if I were once acquitted your sons would pursue what Socrates teaches and all be thoroughly corrupted—if with this in view you were to say to me, "Socrates, we shall not at this time be persuaded by Meletus, and we dismiss you. But on this condition: that you no longer pass time in that inquiry of yours, or pursue philosophy. And if you are again taken doing it, you die." If, as I say, you were to dismiss me on that condition, I would reply that I hold you in friendship and regard, Gentlemen of Athens, but I shall obey the God rather than you, and while I have breath and am able I shall not cease to pursue wisdom or to exhort you, charging any of you I happen to meet in my accustomed manner. "You are the best of men, being an Athenian, citizen of a city honored for wisdom and power beyond all others. Are you then not ashamed to care for the getting of money, and reputation, and public honor, while yet having no thought or concern for truth and understanding and the greatest possible excellence of your soul?" And if some one of you disputes this, and says he does care, I shall not immediately dismiss him and go away. I shall question him and examine him and test him, and if he does not seem to me to possess virtue, and yet says he does, I shall rebuke him for counting of more importance things which by comparison are worthless. I shall do this to young and old, citizen and stranger, whomever I happen to meet, but I shall do it especially to citizens, inasmuch as they are more nearly related to me. For the God commands this, be well assured, and I believe that you have yet to gain in this City a greater good than my service to the God. I go about doing nothing but persuading you, young and old, to care not for body or money in place of, or so much as, excellence of soul. I tell you that virtue does not come from money, but money and all other human goods both public and private from virtue. If in saying this I corrupt the youth, that would be harm indeed. But anyone who claims I say other than this speaks falsehood. In these matters, Gentlemen of Athens, believe Anytus, or do not. Dismiss me, or do not. For I will not do otherwise, even if I am to die for it many times over.

Please do not make a disturbance, Gentlemen. Abide in my request and do not interrupt what I have to say, but listen. Indeed, I think you will benefit by listening. I am going to tell you certain things at which you may perhaps cry out; please do not do it. Be well assured that if you kill me, and if I am the sort of man I claim, you will harm me less than you harm yourselves. There is no harm a Meletus or Anytus can do me; it is not possible, for it does not, I think, accord with divine law that a better man be harmed by a worse. Meletus perhaps can kill me, or exile me, or disenfranchise me; and perhaps he and others too think those things great evils. I do not. I think it a far greater evil to do what he is now doing, attempting to kill a man unjustly. And so, Gentlemen of Athens, I am far from making a defense for my own sake, as some might think; I make it for yours, lest you mistake the gift the God has given you and cast your votes against me. If you kill me, you will not easily find such another man as I, a man who—if I may put it a bit absurdly—has been fastened as it were to the City by the God as to a large and well-bred horse, a horse grown sluggish because of its size, and in need of being roused by a kind of gadfly. Just so, I think, the God has fastened me to the City. I rouse you. I persuade you. I upbraid you. I never stop lighting on each one of you, everywhere, all day long. Such another will not easily come to you again, Gentlemen, and if you are persuaded by me, you will spare me. But perhaps you are angry, as men roused from sleep are angry, and perhaps you will swat me, persuaded by Meletus that you may lightly kill. Then will you continue to sleep out your lives, unless the God sends someone else to look after you.

That I am just that, a gift from the God to the City, you may recognize from this: it scarcely seems a human matter merely, that I should take no thought for anything of my own and endure the neglect of my house and its affairs for these long years now, and ever attend to yours, going to each of you in private like a father or elder brother, persuading you to care for virtue. If I got something from it, if I took pay for this kind of exhortation, that would explain it. But as things are, you can see for yourselves that even my accusers, who have accused me so shamefully of everything else, could not summon shamelessness enough to provide witnesses to testify that I ever took pay or asked for it. For it is enough, I think, to provide my poverty as witness to the truth of what I say.

Perhaps it may seem peculiar that I go about in private advising men and busily inquiring, and yet do not enter your Assembly in public to advise the City. The reason is a thing you have heard me mention many times in many places, that something divine and godlike comes to me—which Meletus, indeed, mocked in his indictment. I have had it from childhood. It comes as a kind of voice, and when it comes, it always turns me away from what I am about to do, but never toward it. That is what opposed my entering political life, and I think it did well to oppose. For be well assured, Gentlemen of Athens, that had I attempted long since to enter political affairs, I should long since have been destroyed—to the benefit of neither you nor myself.

Please do not be angry at me for telling the simple truth. It is impossible for any man to be spared if he legitimately opposes you or any other democratic majority, and prevents many unjust and illegal things from occurring in his city. He who intends to fight for what is just, if he is to be spared even for a little time, must of necessity live a private rather than a public life.

I shall offer you a convincing indication of this—not words, but what you respect, deeds. Hear, then, what befell me, so that you may know that I will not through fear of death give way to any man contrary to what is right, even if I am destroyed for it. I shall tell you a thing which is tedious—it smacks of the law courts—but true. Gentlemen of Athens, I never held other office in the City, but I was once a member of the Council. And it happened that our Tribe, Antiochis, held the Prytanate when you decided to judge as a group the cases of the ten generals who had failed to gather up the bodies of the slain in the naval battle—illegally, as later it seemed to all of you. But at that time, I alone of the Prytanies opposed doing a thing contrary to law, and cast my vote against it. And when the orators were ready to impeach me and have me arrested—you urging them on with your shouts—I thought that with law and justice on my side I must run the risk, rather than concur with you in an unjust decision through fear of bonds or death. Those things happened while the City was still under the Democracy. But when Oligarchy came, the Thirty in turn summoned me along with four others to the Rotunda and ordered us to bring back Leon the Salamanian from Salamis so that he might be executed, just as they ordered many others to do such things, planning to implicate as many people as possible in their own guilt. But I then showed again, not by words but deeds, that death, if I may be rather blunt, was of no concern whatever to me; to do nothing unjust or unholy—that was my concern. Strong as it was, that oligarchy did not so frighten me as to do a thing unjust, and when we departed the Rotunda, the other four went into Salamis and brought back Leon, and I left and went home. I might have been killed for that, if the oligarchy had not shortly afterward been overthrown. And of these things you will have many witnesses.

Now, do you think I would have lived so many years if I had been in public life and acted in a manner worthy of a good man, defending what is just and counting it, as is necessary, of first importance? Far from it, Gentlemen of Athens. Not I, and not any other man. But through my whole life I have shown myself to be that sort of man in public affairs, the few I've engaged in; and I have shown myself the same man in private. I never gave way to anyone contrary to what is just—not to others, and certainly not to those slanderously said to be my pupils. In fact, I have never been teacher to anyone. If, in speaking and tending my own affairs, anyone wished to hear me, young or old, I never begrudged him; nor do I discuss for a fee and not otherwise. To rich and poor alike I offer myself as a questioner, and if anyone wishes to answer, he may hear what I have to say. And if any of them turned out to be useful men, or any did not, I cannot justly be held responsible. To none did I promise instruction, and none did I teach; if anyone says that he learned from me or heard in private what others did not, you may rest assured he is not telling the truth.

Why is it, then, that some people enjoy spending so much time with me? You have heard, Gentlemen of Athens: I told you the whole truth. It is because they enjoy hearing people tested who think they are wise and are not. After all, it is not unamusing. But for my own part, as I say, I have been ordered to do this by God—in oracles, in dreams, in every way in which other divine apportionment ever ordered a man to do anything.

These things, Gentlemen of Athens, are both true and easily tested. For if I am corrupting some of the youth, and have corrupted others, it must surely be that some among them, grown older, if they realize that I counseled them toward evil while young, would now come forward to accuse me and exact a penalty. And if they were unwilling, then some of their relatives—fathers, brothers, other kinsmen—if their own relatives had suffered evil at my hands, would now remember, and exact a penalty. Certainly there are many such men I see present. Here is Crito, first, of my own age and deme, father of Critobulus; then there is Lysanias of Sphettos, father of Aeschines here. Next there is Antiphon of Cephisus, father of Epigenes. Then there are others whose brothers engaged in this pastime. There is Nicostratus, son of Theozotides, brother of Theodotus—and Theodotus is dead, so he could not have swayed him—and Paralus here, son of Demodocus, whose brother was Theages. And here is Adeimantus, son of Ariston, whose brother is Plato there; and Aeantodorus, whose brother is Apollodorus here. I could name many others, some of whom at least Meletus ought certainly have provided in his speech as witnesses. If he forgot it then, let him do it now—I yield the floor—and let him say whether he has any witnesses of the sort. You will find that quite to the contrary, Gentlemen, every one of these men is ready to help me—I, who corrupt their relatives, as Meletus and Anytus claim. Those who are themselves corrupted might perhaps have reason to help me; but their relatives are older men who have not been corrupted. What reason could they have for supporting me except that it is right and just, because they know Meletus is lying and I am telling the truth?

Very well then, Gentlemen. This, and perhaps a few other things like it, is what I have to say in my defense. Perhaps some of you will remember his own conduct and be offended, if when brought to trial on a lesser charge than this, he begged his judges with tearful supplication, and caused his children to come forward so that he might be the more pitied, along with other relatives and a host of friends; whereas I shall do none of these things, even though I am, as it would seem at least, in the extremity of danger. Perhaps someone with this in mind may become hardened against me; angered by it, he may cast his vote in anger. If this is true of any of you—not that I expect it, but if it is—I think it might be appropriate to say, "I too have relatives, my friend; for as Homer puts it, I am not 'of oak and rock,' but born of man, so I have relatives—yes, and sons too, Gentlemen of Athens, three of them, one already a lad and two of them children. Yet not one of them have I caused to come forward here, and I shall not beg you to acquit me." Why not? Not out of stubbornness, Gentlemen of Athens, nor disrespect for you. Whether or not I am confident in the face of death is another story; but I think that my own honor, and yours, and that of the whole City would suffer, if I were to behave in this way, I being of the age I am and having the name I have— truly or falsely, it being thought that Socrates is in some way superior to most men. If those of you reputed to be superior in wisdom or courage or any other virtue whatever are to be men of this sort, it would be disgraceful; I have often seen such people behave surprisingly when put on trial, even though they had a reputation to uphold, because they were persuaded that they would suffer a terrible thing if they were put to death— as though they would be immortal if you did not kill them. I think they cloak the City in shame, so that a stranger might think that those among the Athenians who are superior in virtue, and whom the Athenians themselves judge worthy of office and other honors, are no better than women. These are things, Gentlemen of Athens, which those of you who have a reputation to uphold ought not to do; nor if we defendants do them, ought you permit it. You ought rather make it clear that you would far rather cast your vote against a man who stages these pitiful scenes, and makes the City a butt of mockery, than against a man who shows quiet restraint.

But apart from the matter of reputation, Gentlemen, it does not seem to me just to beg a judge, or to be acquitted by begging; it is rather just to teach and persuade. The judge does not sit to grant justice as a favor, but to render judgment; he has sworn no oath to gratify those whom he sees fit, but to judge according to law. We ought not accustom you, nor ought you become accustomed, to forswear yourselves; it is pious in neither of us. So do not expect me, Gentlemen of Athens, to do such things in your presence as I believe to be neither honorable nor just nor holy, especially since, by Zeus, it is for impiety that I am being prosecuted by this fellow Meletus here. For clearly, if I were to persuade and compel you by supplication, you being sworn as judges, I would teach you then indeed not to believe that there are gods, and in making my defense I would in effect accuse myself of not acknowledging them. But that is far from so; I do acknowledge them, Gentlemen of Athens, as no one of my accusers does, and to you and to the God I now commit my case, to judge in whatever way shall be best both for me and for you.

I am not distressed, Gentlemen of Athens, at what has happened, nor angered that you have cast your votes against me. Many things contribute to this, among them the fact that I expected it. I am much more surprised at the number of votes either way: I did not think it would be by so little, but by more. As it is, it seems, if only thirty votes had fallen otherwise, I would have been acquitted. And so far as Meletus at least is concerned, it seems to me, I am already acquitted—and more than acquitted, since it is clear that if Anytus and Lycon had not come forward to accuse me, Meletus would have been fined a thousand drachmas for not obtaining a fifth part of the vote.

The man demands death for me. Very well. Then what counterpenalty shall I propose to you, Gentlemen of Athens? Clearly something I deserve, but what? What do I deserve to pay or suffer because I did not through life keep quiet, and yet did not concern myself, as the multitude do, with money or property or military and public honors and other office, or the secret societies and political clubs which keep cropping up in the City, believing that I was really too reasonable and temperate a man to enter upon these things and survive. I did not go where I could benefit neither you nor myself; instead, I went to each of you in private, where I might perform the greatest service. I undertook to persuade each of you not to care for anything which belongs to you before first caring for yourselves, nor to care for anything which belongs to the City before caring for the City itself, and so too with everything else. Now, what do I deserve to suffer for being this sort of man? Some good thing, Gentlemen of Athens, if penalty is really to be assessed according to desert. What then is fitting for a poor man who has served his City well, and needs leisure to exhort you? Why, Gentlemen of Athens, nothing is more fitting for such a man than to be fed in the Prytaneum, at the common table of the City—yes, and far more fitting than for one of you who has been an Olympic victor in the single-horse or two- or four-horse chariot races. For he makes you seem happy, whereas I make you happy in truth, and he does not need subsistence, and I do. If then I must propose a penalty I justly deserve, I propose that, public subsistence in the Prytaneum.

Perhaps some of you will think that in saying this I speak much as I spoke of tears and pleading, out of stubborn pride. That is not so, Gentlemen of Athens, though something of this sort is: I am persuaded that I have not intentionally wronged any man, but I cannot persuade you of it; we have talked so short a time. Now, I believe if you had a law, as other people do, that cases involving death shall not be decided in a single day, that you would be persuaded; but as things are, it is not easy in so short a time to do away with slanders grown so great. Being persuaded, however, that I have wronged no one, I am quite unwilling to wrong myself, or claim that I deserve some evil and propose any penalty of the kind. What is there to fear? That I may suffer the penalty Meletus proposes, when as I say, I do not know whether it is good or evil? Shall I choose instead a penalty I know very well to be evil? Imprisonment, perhaps? But why should I live in prison, a slave to men who happen to occupy office as the Eleven? A fine, then, and imprisonment till I pay it? But that comes to the same thing, since I have no money to pay it. Shall I then propose exile? Perhaps you would accept that. But I must indeed love life and cling to it dearly, Gentlemen, if I were so foolish as to think that, although you, my own fellow-citizens, cannot bear my pursuits and discussions, which have become so burdensome and hateful that you now seek to be rid of them, others will bear them lightly. No, Gentlemen. My life would be fine indeed, if at my age I went to live in exile, always moving from city to city, always driven out. For be well assured that wherever I go, the young men will listen to what I say as they do here; if I turn them away they will themselves drive me out, appealing to their elders; if I do not turn them away, their fathers and relations will drive me out in their behalf.

Perhaps someone may say, "Would it not be possible for you to live in exile, Socrates, if you were silent and kept quiet?" But this is the hardest thing of all to make some of you believe. If I say that to do so would be to disobey the God, and therefore I cannot do it, you will not believe me because you will think that I am being sly and dishonest. If on the other hand I say that the greatest good for man is to fashion arguments each day about virtue and the other things you hear me discussing, when I examine myself and others, and that the unexamined life is not for man worth living, you will believe what I say still less.

I claim these things are so, Gentlemen; but it is not easy to convince you. At the same time, I am not accustomed to think myself deserving of any evil. If I had money, I would propose a fine as great as I could pay—there would be no harm in that. But as things stand, I have no money, unless the amount I can pay is the amount you are willing to exact of me. I might perhaps be able to pay a mina of silver. So I propose a penalty in that amount. But Plato here, Gentlemen of Athens, and Crito and Critobulus and Apollodorus bid me propose thirty minas, and they will stand surety. So I propose that amount. You have guarantors sufficient for the sum.

For the sake of only a little time, Gentlemen of Athens, you are to be accused by those who wish to revile the City of having killed Socrates, a wise man—for those who wish to reproach you will say I am wise even if I am not. And if you had only waited a little, the thing would have come of its own initiative. You see my age. You see how far spent my life already is, how near to death.

I say this, not to all of you, but to those of you who voted to condemn me. To them I also say this. Perhaps you think, Gentlemen of Athens, that I have been convicted for lack of words to persuade you, had I thought it right to do and say anything to be acquitted. Not so. It is true I have been convicted for a lack; not a lack of words, but lack of bold shamelessness, unwillingness to say the things you would find it pleasant to hear—weeping and wailing, saying and doing many things I claim to be unworthy of me, but things of the sort you are accustomed to hear from others. I did not then think it necessary to do anything unworthy of a free man because of danger; I do not now regret so having conducted my defense; and I would far rather die with that defense than live with the other. Neither in court of law nor in war ought I or any man contrive to escape death by any means possible. Often in battle it becomes clear that a man may escape death by throwing down his arms and turning in supplication to his pursuers; and there are many other devices for each of war's dangers, so that one can avoid dying if he is bold enough to say and do anything at all. It is not difficult to escape death, Gentlemen; it is more difficult to escape wickedness, for wickedness runs faster than death. And now I am old and slow, and I have been caught by the slower runner. But my accusers are clever and quick, and they have been caught by the faster runner, namely Evil. I now take my leave, sentenced by you to death; they depart convicted by Truth for injustice and wickedness. I abide in my penalty, and they in theirs. That is no doubt as it should be, and I think it is fit.

I desire next to prophesy to you who condemned me. For I have now reached that point where men are especially prophetic, when they are about to die. I say to you who have decreed my death that to you there will come hard on my dying a thing far more difficult to bear than the death you have visited upon me. You have done this thing in the belief that you would be released from submitting to examination in your lives. I say that it will turn out otherwise. Those who come to examine you will be more numerous, and I have up to now restrained them, though you perceived it not. They will be more harsh inasmuch as they are younger, and you will be the more troubled. If you think by killing to hold back the reproach due you for not living rightly, you are profoundly mistaken. That release is neither possible nor honorable. The release which is both most honorable and most easy is not to cut down others, but to so prepare yourselves that you will be as good as possible. This I utter as prophecy to you who voted for my condemnation, and take my leave.

But with you who voted for my acquittal, I should be glad to discuss the nature of what has happened, now, while the authorities are busy and I am not yet gone where, going, I must die. Abide with me, Gentlemen, this space of time; nothing prevents our talking with each other while we still can. To you, as my friends, I wish to display the meaning of what has now fallen to my lot. A remarkable thing has occurred, Gentlemen and Judges—and I am correct in calling you Judges. My accustomed oracle, which is divine, always came quite frequently before in everything, opposing me even in trivial matters if I was about to err. And now a thing has fallen to my lot which you also see yourselves, a thing which some might think, and do in fact believe, to be ultimate among evils. But the sign of the God did not oppose me early this morning when I left my house, nor when I came up to the courtroom here, nor at any point in my argument in anything I was about to say. And yet in many places, in other arguments, it has checked me right in the middle of speaking; but today it has not opposed me in any way, in none of my deeds, in none of my words. What do I take to be the reason? I will tell you. Very likely what has fallen to me is good, and those among us who think that death is an evil are wrong. There has been convincing indication of this. For the accustomed sign would surely have opposed me, if I were not in some way acting for good.

Let us also consider a further reason for high hope that death is good. Death is one of two things. Either to be dead is not to exist, to have no awareness at all, or it is, as the stories tell, a kind of alteration, a change of abode for the soul from this place to another. And if it is to have no awareness, like a sleep when the sleeper sees no dream, death would be a wonderful gain; for I suppose if someone had to pick out that night in which he slept and saw no dream, and put the other days and nights of his life beside it, and had to say after inspecting them how many days and nights he had lived in his life which were better and sweeter, I think that not only any ordinary person but even the Great King himself would find them easily numbered in relation to other days, and other nights. If death is that, I say it is gain; for the whole of time then turns out to last no longer than a single night. But if on the contrary death is like taking a journey, passing from here to another place, and the stories told are true, and all who have died are there—what greater good might there be, my Judges? For if a man once goes to the place of the dead, and takes leave of those who claim to be judges here, he will find the true judges who are said to sit in judgment there—Minos, Rhadamanthus, Aeacus, Triptolemus, and the other demigods and heroes who lived just lives. Would that journey be worthless? And again, to meet Orpheus and Musaeus, Hesiod and Homer—how much would any of you give? I at least would be willing to die many times over, if these things are true. I would find a wonderful pursuit there, when I met Palamedes, and Ajax, son of Telemon, and any others among the ancients done to death by unjust verdicts, and compared my experiences with theirs. It would not, I think, be unamusing. But the greatest thing, surely, would be to test and question there as I did here—who among them is wise? Who thinks he is and is not? How much might one give, my Judges, to examine the man who led the great army against Troy, or Odysseus, or Sisyphus, or a thousand other men and women one might mention—to converse with them, to associate with them, to examine them—why, it would be inconceivable happiness. Especially since they surely do not kill you for it there. For they are happier there than men are here in other ways, and they are already immortal for the rest of time, if the stories told are true.

But you too, my Judges, must be of good hope concerning death. You must recognize that this one thing is true: that there is no evil for a good man either in living or in dying, and that the gods do not neglect his affairs. What has now come to me did not come of its own initiative. It is clear to me that to die now and be released from my affairs is better for me. That is why the sign did not turn me back, and I bear no anger whatever toward those who voted to condemn me, or toward my accusers. And yet, it was not with this in mind that they accused and convicted me. They thought to do harm, and for that they deserve blame. But this much would I ask of them: when my sons are grown, Gentlemen, exact a penalty of them: give pain to them exactly as I gave pain to you, if it seems to you that they care more for wealth or anything else than they care for virtue. And if they seem to be something and are nothing, rebuke them as I rebuked you, because they do not care for what they ought, because they think themselves something and are worth nothing. And should you do that, both I and my sons will have been justly dealt with at your hands.

But it is already the hour of parting—I to die and you to live. Which of us goes to the better is unclear to all but the God.

JOSEPH CONRAD was born in 1857 near Kiev, Russia
to Polish parents who christened him Jozef Tedor
Korzeniowski. His father, a poet and translator, was
active with political groups attempting to free Poland
from Russian domination, and as a result the family
was exiled by the czarist government to northern Russia
in 1861. It was during their exile that his father introduced Conrad to English literature, which he was trans-
lating. Conrad's mother died in 1865 and his father
in 1869, both of tuberculosis. Conrad was then cared
for by an uncle, who sent him to school in Poland and
Switzerland. Formal education was not to Conrad's
liking, and at seventeen he enlisted as a sailor in the
French merchant marine. Conrad lived a sailor's life for
twenty years, sailing mostly on English ships. He studied
the English language and became an English subject.
While waiting in London for work aboard a ship in
1889, Conrad began writing *Almayer's Folly* (1895), his
first novel. He gave up sailing in 1894 to write addi-
tional novels and short stories. Conrad died in England
in 1924.

From *Heart of Darkness*. Publisher: Penguin Books,
1973.

Heart of Darkness

1

The *Nellie,* a cruising yawl, swung to her anchor without a flutter of the sails, and was at rest. The flood had made, the wind was nearly calm, and being bound down the river, the only thing for it was to come to and wait for the turn of the tide.

The sea-reach of the Thames stretched before us like the beginning of an interminable waterway. In the offing the sea and the sky were welded together without a joint, and in the luminous space the tanned sails of the barges drifting up with the tide seemed to stand still in red clusters of canvas sharply peaked, with gleams of varnished sprits. A haze rested on the low shores that ran out to sea in vanishing flatness. The air was dark above Gravesend, and farther back still seemed condensed into a mournful gloom, brooding motionless over the biggest, and the greatest, town on earth.

The Director of Companies was our captain and our host. We four affectionately watched his back as he stood in the bows looking to seaward. On the whole river there was nothing that looked half so nautical. He resembled a pilot, which to a seaman is trustworthiness personified. It was difficult to realize his work was not out there in the luminous estuary, but behind him, within the brooding gloom.

Between us there was, as I have already said somewhere, the bond of the sea. Besides holding our hearts together through long periods of separation, it had the effect of making us tolerant of each other's yarns—and even convictions. The Lawyer—the best of old fellows—had, because of his many years and many virtues, the only cushion on deck, and was lying on the only rug. The Accountant had brought out already a box of dominoes, and was toying architecturally with the bones. Marlow sat cross-legged right aft, leaning against the mizzen-mast. He had sunken cheeks, a yellow complexion, a straight back, an ascetic aspect, and, with his arms dropped, the palms of hands outwards, resembled an idol. The director, satisfied the anchor had good hold, made his way aft and sat down amongst us. We exchanged a few words lazily. Afterwards there was silence on board the yacht. For some reason or other we did not begin that game of dominoes. We felt meditative, and fit for nothing but placid staring. The day was ending in a serenity of still and exquisite brilliance. The water shone pacifically; the sky, without a speck, was a benign immensity of unstained light; the very mist on the Essex marshes was like a gauzy and radiant fabric, hung from the wooded rises inland, and draping the low shores in diaphanous folds. Only the gloom to the west, brooding over the upper reaches, became more sombre every minute, as if angered by the approach of the sun.

And at last, in its curved and imperceptible fall, the sun sank low, and from glowing white changed to a dull red without rays and without heat, as if about to go out suddenly, stricken to death by the touch of that gloom brooding over a crowd of men.

Forthwith a change came over the waters, and the serenity became less brilliant but more profound. The old river in its broad reach rested unruffled at the decline of day, after ages of good service done to the race that peopled its banks, spread out in the tranquil dignity of a waterway leading to the uttermost ends of the earth. We looked at the venerable stream not in the vivid flush of a short day that comes and departs for ever, but in the august light of abiding memories. And indeed nothing is easier for a man who has, as the phrase goes, "followed the sea" with reverence and affection, than to evoke the great spirit of the past upon the lower reaches of the Thames. The tidal current runs to and fro in its unceasing service, crowded with memories of men and ships it had borne to the rest of home or to the battles of the sea. It had known and served all the men of whom the nation is proud, from Sir Francis Drake to Sir John Franklin, knights all, titled and untitled—the great knights-errant of the sea. It had borne all the ships whose names are like jewels flashing in the night of time, from the *Golden Hind* returning with her round flanks full of treasure, to be visited by the Queen's Highness and thus pass out of the gigantic tale, to the *Erebus* and *Terror,* bound on other conquests—and that never returned. It had known the ships and the men. They had sailed from Deptford, from Greenwich, from Erith—the adventurers and the settlers; kings' ships and the ships of men on 'Change; captains, admirals, the dark "interlopers" of the Eastern trade, and the commissioned "generals" of East India fleets. Hunters for gold or pursuers of fame, they all had gone out on that stream, bearing the sword, and often the torch, messengers of the might within the land, bearers of a spark from the sacred fire. What greatness had not floated on the ebb of that river into the mystery of an unknown earth! . . . The dreams of men, the seed of commonwealths, the germs of empires.

The sun set; the dusk fell on the stream, and lights began to appear along the shore. The Chapman lighthouse, a three-legged thing erect on a mud-flat, shone strongly. Lights of ships moved in the fairway—a great stir of lights going up and going down. And farther west on the upper reaches the place of the monstrous town was still marked ominously on the sky, a brooding gloom in sunshine, a lurid glare under the stars.

"And this also," said Marlow suddenly, "has been one of the dark places of the earth."

He was the only man of us who still "followed the sea." The

worst that could be said of him was that he did not represent
his class. He was a seaman, but he was a wanderer, too, while
most seamen lead, if one may so express it, a sedentary life.
Their minds are of the stay-at-home order, and their home is
always with them—the ship; and so is their country—the sea.
One ship is very much like another, and the sea is always the
same. In the immutability of their surroundings the foreign
shores, the foreign faces, the changing immensity of life, glide
past, veiled not by a sense of mystery but by a slightly disdainful
ignorance; for there is nothing mysterious to a seaman unless it
be the sea itself, which is the mistress of his existence and as
inscrutable as Destiny. For the rest, after his hours of work, a
casual stroll or a casual spree on shore suffices to unfold for him
the secret of a whole continent, and generally he finds the secret
not worth knowing. The yarns of seamen have a direct simplicity,
the whole meaning of which lies within the shell of a cracked
nut. But Marlow was not typical (if his propensity to spin yarns
be excepted), and to him the meaning of an episode was not
inside like a kernel but outside, enveloping the tale which brought
it out only as a glow brings out a haze, in the likeness of one
of these misty halos that sometimes are made visible by the
spectral illumination of moonshine.

His remark did not seem at all surprising. It was just like
Marlow. It was accepted in silence. No one took the trouble to
grunt even; and presently he said, very slow—

"I was thinking of very old times, when the Romans first
came here, nineteen hundred years ago—the other day . . . Light
came out of this river since—you say Knights? Yes; but it is
like a running blaze on a plain, like a flash of lightning in the
clouds. We live in the flicker—may it last as long as the old
earth keeps rolling! But darkness was here yesterday. Imagine
the feelings of a commander of a fine—what d'ye call 'em?—
trireme in the Mediterranean, ordered suddenly to the north;
run overland across the Gauls in a hurry; put in charge of one

of these craft the legionaries—a wonderful lot of handy men they must have been, too—used to build, apparently by the hundred, in a month or two, if we may believe what we read. Imagine him here—the very end of the world, a sea the colour of lead, a sky the colour of smoke, a kind of ship about as rigid as a concertina—and going up this river with stores, or orders, or what you like. Sand-banks, marshes, forests, savages—precious little to eat fit for a civilized man, nothing but Thames water to drink. No Falernian wine here, no going ashore. Here and there a military camp lost in a wilderness, like a needle in a bundle of hay—cold, fog, tempests, disease, exile, and death—death skulking in the air, in the water, in the bush. They must have been dying like flies here. Oh, yes—he did it. Did it very well, too, no doubt, and without thinking much about it either, except afterwards to brag of what he had done through his time, perhaps. They were men enough to face the darkness. And perhaps he was cheered by keeping his eye on a chance of promotion to the fleet at Ravenna by-and-by, if he had good friends in Rome and survived the awful climate. Or think of a decent young citizen in a toga—perhaps too much dice, you know—coming out here in the train of some prefect, or tax-gatherer, or trader even, to mend his fortunes. Land in a swamp, march through the woods, and in some inland post feel the savagery, the utter savagery, had closed round him— all that mysterious life of the wilderness that stirs in the forest, in the jungles, in the hearts of wild men. There's no initiation either into such mysteries. He has to live in the midst of the incomprehensible, which is also detestable. And it has a fascination, too, that goes to work upon him. The fascination of the abomination—you know, imagine the growing regrets, the longing to escape, the powerless disgust, the surrender, the hate."

He paused.

"Mind," he began again, lifting one arm from the elbow, the palm of the hand outwards, so that, with his legs folded before him, he had the pose of a Buddha preaching in European clothes and without a lotus-flower—"Mind, none of us would feel exactly like this. What saves us is efficiency—the devotion to efficiency. But these chaps were not much account, really. They were no colonists; their administration was merely a squeeze, and nothing more, I suspect. They were conquerors, and for that you want only brute force—nothing to boast of, when you have it, since your strength is just an accident arising from the weakness of others. They grabbed what they could get for the sake of what was to be got. It was just robbery with violence, aggravated murder on a great scale, and men going at it blind—as is very proper for those who tackle a darkness. The conquest of the earth, which mostly means the taking it away from those who have a different complexion or slightly flatter noses than ourselves, is not a pretty thing when you look into it too much. What redeems it is the idea only. An idea at the back of it; not a sentimental pretence but an idea; and an unselfish belief in the idea—something you can set up, and bow down before, and offer a sacrifice to . . ."

He broke off. Flames glided in the river, small green flames, red flames, white flames, pursuing, overtaking, joining, crossing each other—then separating slowly or hastily. The traffic of the great city went on in the deepening night upon the sleepless river. We looked on, waiting patiently—there was nothing else to do till the end of the flood; but it was only after a long silence, when he said, in a hesitating voice, "I suppose you fellows remember I did once turn fresh-water sailor for a bit," that we knew we were fated, before the ebb began to run, to hear about one of Marlow's inconclusive experiences.

"I don't want to bother you much with what happened to me personally," he began, showing in this remark the weakness of many tellers of tales who seem so often unaware of what their audience would best like to hear; "yet to understand the effect of it on me you ought to know how I got out there, what I

saw, how I went up that river to the place where I first met
the poor chap. It was the farthest point of navigation and the
culminating point of my experience. It seemed somehow to
throw a kind of light on everything about me—and into my
thoughts. It was sombre enough, too—and pitiful—not extraordinary in any way—not very clear either. No, not very clear.
And yet it seemed to throw a kind of light.

"I had then, as you remember, just returned to London after
a lot of Indian Ocean, Pacific, China Seas—a regular dose of
the East—six years or so, and I was loafing about, hindering
you fellows in your work and invading your homes, just as
though I had got a heavenly mission to civilize you. It was very
fine for a time, but after a bit I did get tired of resting. Then
I began to look for a ship—I should think the hardest work on
earth. But the ships wouldn't even look at me. And I got tired
of that game, too.

"Now when I was a little chap I had a passion for maps. I
would look for hours at South America, or Africa, or Australia,
and lose myself in all the glories of exploration. At that time
there were many blank spaces on the earth, and when I saw one
that looked particularly inviting on a map (but they all look
that) I would put my finger on it and say, When I grow up I
will go there. The North Pole was one of these places, I remember. Well, I haven't been there yet, and shall not try now.
The glamour's off. Other places were scattered about the Equa-
tor, and in every sort of latitude all over the two hemispheres.
I have been in some of them, and . . . well, we won't talk about
that. But there was one yet—the biggest, the most blank, so to
speak—that I had a hankering after.

"True, by this time it was not a blank space any more. It
had got filled since my boyhood with rivers and lakes and names.
It had ceased to be a blank space of delightful mystery—a white
patch for a boy to dream gloriously over. It had become a place
of darkness. But there was in it one river especially, a mighty

big river, that you could see on the map, resembling an immense
snake uncoiled, with its head in the sea, its body at rest curving
afar over a vast country, and its tail lost in the depths of the
land. And as I looked at the map of it in a shop-window, it
fascinated me as a snake would a bird—a silly little bird. Then
I remembered there was a big concern, a Company for trade on
that river. Dash it all! I thought to myself, they can't trade
without using some kind of craft on that lot of fresh water—
steamboats! Why shouldn't I try to get charge of one? I went
on along Fleet Street, but could not shake off the idea. The
snake had charmed me.

"You understand it was a Continental concern, that Trading
society; but I have a lot of relations living on the Continent,
because it's cheap and not so nasty as it looks, they say.

"I am sorry to own I began to worry them. This was already
a fresh departure for me. I was not used to get things that way,
you know. I always went my own road and on my own legs
where I had a mind to go. I wouldn't have believed it of myself;
but, then—you see—I felt somehow I must get there by hook
or by crook. So I worried them. The men said, 'My dear fellow,'
and did nothing. Then—would you believe it?—I tried the
women. I, Charlie Marlow, set the women to work—to get a
job. Heavens! Well, you see, the notion drove me. I had an
aunt, a dear enthusiastic soul. She wrote: 'It will be delightful.
I am ready to do anything, anything for you. It is a glorious
idea. I know the wife of a very high personage in the Administration, and also a man who has lots of influence with,' etc.,
etc. She was determined to make no end of fuss to get me
appointed skipper of a river steamboat, if such was my fancy.

"I got my appointment—of course; and I got it very quick.
It appears the Company had received news that one of their
captains had been killed in a scuffle with the natives. This was
my chance, and it made me the more anxious to go. It was
only months and months afterwards, when I made the attempt

to recover what was left of the body, that I heard the original quarrel arose from a misunderstanding about some hens. Yes, two black hens. Fresleven—that was the fellow's name, a Dane—thought himself wronged somehow in the bargain, so he went ashore and started to hammer the chief of the village with a stick. Oh, it didn't surprise me in the least to hear this, and at the same time to be told that Fresleven was the gentlest, quietest creature that ever walked on two legs. No doubt he was; but he had been a couple of years already out there engaged in the noble cause, you know, and he probably felt the need at last of asserting his self-respect in some way. Therefore he whacked the old nigger mercilessly, while a big crowd of his people watched him, thunderstruck, till some man—I was told the chief's son—in desperation at hearing the old chap yell, made a tentative jab with a spear at the white man—and of course it went quite easy between the shoulder-blades. Then the whole population cleared into the forest, expecting all kinds of calamities to happen, while, on the other hand, the steamer Fresleven commanded left also in a bad panic, in charge of the engineer, I believe. Afterwards nobody seemed to trouble much about Fresleven's remains, till I got out and stepped into his shoes. I couldn't let it rest, though; but when an opportunity offered at last to meet my predecessor, the grass growing through his ribs was tall enough to hide his bones. They were all there. The supernatural being had not been touched after he fell. And the village was deserted, the huts gaped black, rotting, all askew within the fallen enclosures. A calamity had come to it, sure enough. The people had vanished. Mad terror had scattered them, men, women, and children, through the bush, and they had never returned. What became of the hens I don't know either. I should think the cause of progress got them, anyhow. However, through this glorious affair I got my appointment, before I had fairly begun to hope for it.

"I flew around like mad to get ready, and before forty-eight

hours I was crossing the Channel to show myself to my em-
ployers, and sign the contract. In a very few hours I arrived in
a city that always makes me think of a whited sepulchre. Prejudice no doubt. I had no difficulty in finding the Company's
offices. It was the biggest thing in the town, and everybody I
met was full of it. They were going to run an over-sea empire,
and make no end of coin by trade.

"A narrow and deserted street in deep shadow, high houses,
innumerable windows with venetian blinds, a dead silence, grass
sprouting between the stones, imposing carriage archways right
and left, immense double doors standing ponderously ajar. I
slipped through one of these cracks, went up a swept and un-
garnished staircase, as arid as a desert, and opened the first door
I came to. Two women, one fat and the other slim, sat on straw-
bottomed chairs, knitting black wool. The slim one got up and
walked straight at me—still knitting with downcast eyes—and
only just as I began to think of getting out of her way, as you
would for a somnambulist, stood still, and looked up. Her dress
was as plain as an umbrella-cover, and she turned round without
a word and preceded me into a waiting-room. I gave my name,
and looked about. Deal table in the middle, plain chairs all
round the walls, on one end a large shining map, marked with
all the colours of a rainbow. There was a vast amount of red—
good to see at any time, because one knows that some real work
is done in there, a deuce of a lot of blue, a little green, smears
of orange, and, on the East Coast, a purple patch, to show where
the jolly pioneers of progress drink the jolly lager-beer. However,
I wasn't going into any of these. I was going into the yellow.
Dead in the centre. And the river was there—fascinating—
deadly—like a snake. Ough! A door opened, a white-haired
secretarial head, but wearing a compassionate expression, appeared, and a skinny forefinger beckoned me into the sanctuary.
Its light was dim, and a heavy writing-desk squatted in the

middle. From behind that structure came out an impression of pale plumpness in a frock-coat. The great man himself. He was five feet six, I should judge, and had his grip on the handle-end of ever so many millions. He shook hands, I fancy, murmured vaguely, was satisfied with my French. *Bon voyage.*

"In about forty-five seconds I found myself again in the waiting-room with the compassionate secretary, who, full of desolation and sympathy, made me sign some document. I believe I undertook amongst other things not to disclose any trade secrets. Well, I am not going to.

"I began to feel slightly uneasy. You know I am not used to such ceremonies, and there was something ominous in the atmosphere. It was just as though I had been let into some conspiracy—I don't know—something not quite right; and I was glad to get out. In the outer room the two women knitted black wool feverishly. People were arriving, and the younger one was walking back and forth introducing them. The old one sat on her chair. Her flat cloth slippers were propped up on a foot-warmer, and a cat reposed on her lap. She wore a starched white affair on her head, had a wart on one cheek, and silver-rimmed spectacles hung on the tip of her nose. She glanced at me above the glasses. The swift and indifferent placidity of that look troubled me. Two youths with foolish and cheery countenances were being piloted over, and she threw at them the same quick glance of unconcerned wisdom. She seemed to know all about them and about me, too. An eerie feeling came over me. She seemed uncanny and fateful. Often far away there I thought of these two, guarding the door of Darkness, knitting black wool as for a warm pall, one introducing, introducing continuously to the unknown, the other scrutinizing the cheery and foolish faces with unconcerned old eyes. *Ave!* Old knitter of black wool. *Morituri te salutant.* Not many of those she looked at ever saw her again—not half, by a long way.

"There was yet a visit to the doctor. 'A simple formality,' assured me the secretary, with an air of taking an immense part in all my sorrows. Accordingly a young chap wearing his hat over the left eyebrow, some clerk I suppose—there must have been clerks in the business, though the house was as still as a house in a city of the dead—came from somewhere upstairs, and led me forth. He was shabby and careless, with ink-stains on the sleeves of his jacket, and his cravat was large and billowy, under a chin shaped like the toe of an old boot. It was a little too early for the doctor, so I proposed a drink, and thereupon he developed a vein of joviality. As we sat over our vermouths he glorified the Company's business, and by-and-by I expressed casually my surprise at him not going out there. He became very cool and collected all at once. 'I am not such a fool as I look, quoth Plato to his disciples,' he said sententiously, emptied his glass with great resolution, and we rose.

"The old doctor felt my pulse, evidently thinking of something else the while. 'Good, good for there,' he mumbled, and then with a certain eagerness asked me whether I would let him measure my head. Rather surprised, I said Yes, when he produced a thing like calipers and got the dimensions back and front and every way, taking notes carefully. He was an unshaven little man in a threadbare coat like a gaberdine, with his feet in slippers, and I thought him a harmless fool. 'I always ask leave, in the interests of science, to measure the crania of those going out there,' he said. 'And when they come back, too?' I asked. 'Oh, I never see them,' he remarked; 'and, moreover, the changes take place inside, you know.' He smiled, as if at some quiet joke. 'So you are going out there. Famous. Interesting, too.' He gave me a searching glance, and made another note. 'Ever any madness in your family?' he asked, in a matter-of-fact tone. I felt very annoyed. 'Is that question in the interests of science, too?' 'It would be,' he said, without taking notice of my irritation, 'interesting for science to watch the mental changes

of individuals, on the spot, but . . .' 'Are you an alienist?' I
interrupted. 'Every doctor should be—a little,' answered that
original, imperturbably. 'I have a little theory which you Messieurs who go out there must help me to prove. This is my
share in the advantages my country shall reap from the possession
of such a magnificent dependency. The mere wealth I leave to
others. Pardon my questions, but you are the first Englishman
coming under my observation . . .' I hastened to assure him I
was not in the least typical. 'If I were,' said I, 'I wouldn't be
talking like this with you.' 'What you say is rather profound,
and probably erroneous,' he said, with a laugh. 'Avoid irritation
more than exposure to the sun. Adieu. How do you English
say, eh? Good-bye. Ah! Good-bye. Adieu. In the tropics one
must before everything keep calm.' . . . He lifted a warning
forefinger . . . '*Du calme, du calme. Adieu.*'

"One thing more remained to do—say good-bye to my excellent aunt. I found her triumphant. I had a cup of tea—the
last decent cup of tea for many days—and in a room that most
soothingly looked just as you would expect a lady's drawing-
room to look, we had a long quiet chat by the fireside. In the
course of these confidences it became quite plain to me I had
been represented to the wife of the high dignitary, and goodness
knows to how many more people besides, as an exceptional and
gifted creature—a piece of good fortune for the Company—a
man you don't get hold of every day. Good heavens! and I was
going to take charge of a two-penny-half-penny river steamboat
with a penny whistle attached! It appeared, however, I was also
one of the Workers, with a capital—you know. Something like
an emissary of light, something like a lower sort of apostle.
There had been a lot of such rot let loose in print and talk just
about that time, and the excellent woman, living right in the
rush of all that humbug, got carried off her feet. She talked
about 'weaning those ignorant millions from their horrid ways,'
till, upon my word, she made me quite uncomfortable. I ven-
tured to hint that the Company was run for profit.

" 'You forget, dear Charlie, that the labourer is worthy of his hire,' she said, brightly. It's queer how out of touch with truth women are. They live in a world of their own, and there had never been anything like it, and never can be. It is too beautiful altogether, and if they were to set it up it would go to pieces before the first sunset. Some confounded fact we men have been living contentedly with ever since the day of creation would start up and knock the whole thing over.

"After this I got embraced, told to wear flannel, be sure to write often, and so on—and I left. In the street—I don't know why—a queer feeling came to me that I was an impostor. Odd thing that I, who used to clear out for any part of the world at twenty-four hours' notice, with less thought than most men give to the crossing of a street, had a moment—I won't say of hesitation, but of startled pause, before this commonplace affair. The best way I can explain it to you is by saying that, for a second or two, I felt as though, instead of going to the centre of a continent, I were about to set off for the centre of the earth.

"I left in a French steamer, and she called in every blamed port they have out there, for, as far as I could see, the sole purpose of landing soldiers and custom-house officers. I watched the coast. Watching a coast as it slips by the ship is like thinking about an enigma. There it is before you—smiling, frowning, inviting, grand, mean, insipid, or savage, and always mute with an air of whispering, Come and find out. This one was almost featureless, as if still in the making, with an aspect of monotonous grimness. The edge of a colossal jungle, so dark-green as to be almost black, fringed with white surf, ran straight, like a ruled line, far, far away along a blue sea whose glitter was blurred by a creeping mist. The sun was fierce, the land seemed to glisten and drip with steam. Here and there greyish-whitish specks showed up clustered inside the white surf, with a flag flying above them perhaps. Settlements some centuries old, and still no bigger than pinheads on the untouched expanse of their background. We pounded along, stopped, landed soldiers; went on, landed custom-house clerks to levy toll in what looked like a God-forsaken wilderness, with a tin shed and a flag-pole lost in it; landed more soldiers—to take care of the custom-house clerks, presumably. Some, I heard, got drowned in the surf; but whether they did or not, nobody seemed particularly to care. They were just flung out there, and on we went. Every day the coast looked the same, as though we had not moved; but we passed various places—trading places—with names like Gran' Bassam, Little Popo; names that seemed to belong to some sordid farce acted in front of a sinister back-cloth. The idleness of a passenger, my isolation amongst all these men with whom I had no point of contact, the oily and languid sea, the uniform sombreness of the coast, seemed to keep me away from the truth of things, within the toil of a mournful and senseless delusion. The voice of the surf now and then was a positive pleasure, like the speech of a brother. It was something natural, that had its reason, that had a meaning. Now and then a boat from the shore gave one a momentary contact with reality. It was paddled by black fellows. You could see from afar the white of their eyeballs glistening. They shouted, sang; their bodies streamed with perspiration; they had faces like grotesque masks—these chaps; but they had bone, muscle, a wild vitality, an intense energy of movement, that was as natural and true as the surf along their coast. They wanted no excuse for being there. They were a great comfort to look at. For a time I would feel I belonged still to a world of straight-forward facts; but the feeling would not last long. Something would turn up to scare it away. Once, I remember, we came upon a man-of-war anchored off the coast. There wasn't even a shed there, and she was shelling the bush. It appears the French had one of their wars going on thereabouts. Her ensign dropped limp like a rag; the muzzles of the long six-inch guns stuck out all over the low hull; the greasy, slimy swell swung her up lazily and let her down, swaying her thin masts. In the empty immensity of earth, sky, and water, there she was, incomprehensible, firing into a continent. Pop, would go one of the six-inch guns; a small flame would dart and vanish, a little white smoke would disappear, a tiny projectile would give a feeble screech—and nothing happened. Nothing could happen. There was a touch of insanity in the proceeding, a sense of lugubrious drollery in the sight; and it was not dissipated by somebody on board assuring me earnestly there was a camp of natives—he called them enemies!—hidden out of sight somewhere.

"We gave her her letters (I heard the men in that lonely ship were dying of fever at the rate of three-a-day) and went on. We called at some more places with farcical names, where the merry dance of death and trade goes on in a still and earthy atmosphere as of an overheated catacomb; all along the formless coast bordered by dangerous surf, as if Nature herself had tried to ward off intruders; in and out of rivers, streams of death in life, whose banks were rotting into mud, whose waters, thickened into slime, invaded the contorted mangroves, that seemed to writhe at us in the extremity of an impotent despair. Nowhere did we stop long enough to get a particularized impression, but the general sense of vague and oppressive wonder grew upon me. It was like a weary pilgrimage amongst hints for nightmares.

"It was upward of thirty days before I saw the mouth of the big river. We anchored off the seat of the government. But my work would not begin till some two hundred miles farther on. So as soon as I could I made a start for a place thirty miles higher up.

"I had my passage on a little sea-going steamer. Her captain was a Swede, and knowing me for a seaman, invited me on the bridge. He was a young man, lean, fair, and morose, with lanky hair and a shuffling gait. As we left the miserable little wharf, he tossed his head contemptuously at the shore. 'Been living there?' he asked. I said, 'Yes.' 'Fine lot these government chaps—are they not?' he went on, speaking English with great precision and considerable bitterness. 'It is funny what some people will do for a few francs-a-month. I wonder what becomes of that kind when it goes up country?' I said to him I expected to see that soon. 'So-o-o!' he exclaimed. He shuffled athwart, keeping one eye ahead vigilantly. 'Don't be too sure,' he continued. 'The other day I took up a man who hanged himself on the road. He was a Swede, too.' 'Hanged himself! Why, in God's name?' I cried. He kept on looking out watchfully. 'Who knows? The sun too much for him, or the country perhaps.'

"At last we opened a reach. A rocky cliff appeared, mounds of turned-up earth by the shore, houses on a hill, others with iron roofs, amongst a waste of excavations, or hanging to the declivity. A continuous noise of the rapids above hovered over this scene of inhabited devastation. A lot of people, mostly black and naked, moved about like ants. A jetty projected into the river. A blinding sunlight drowned all this at times in a sudden recrudescence of glare. 'There's your Company's station,' said the Swede, pointing to three wooden barrack-like structures on the rocky slope. 'I will send your things up. Four boxes did you say? So. Farewell.'

"I came upon a boiler wallowing in the grass, then found a path leading up the hill. It turned aside for the boulders, and also for an undersized railway-truck lying there on its back with its wheels in the air. One was off. The thing looked as dead as the carcass of some animal. I came upon more pieces of decaying machinery, a stack of rusty rails. To the left a clump of trees made a shady spot, where dark things seemed to stir feebly. I blinked, the path was steep. A horn tooted to the right, and I saw the black people run. A heavy and dull detonation shook the ground, a puff of smoke came out of the cliff, and that was all. No change appeared on the face of the rock. They were building a railway. The cliff was not in the way or anything; but this objectless blasting was all the work going on.

"A slight clinking behind me made me turn my head. Six black men advanced in a file, toiling up the path. They walked erect and slow, balancing small baskets full of earth on their heads, and the clink kept time with their footsteps. Black rags were wound round their loins, and the short ends behind waggled to and fro like tails. I could see every rib, the joints of their limbs were like knots in a rope; each had an iron collar on his neck, and all were connected together with a chain whose bights swung between them, rhythmically clinking. Another report from the cliff made me think suddenly of that ship of war I had seen firing into a continent. It was the same kind of ominous voice; but these men could by no stretch of imagination be called enemies. They were called criminals, and the outraged law, like the bursting shells, had come to them, an insoluble mystery from the sea. All their meagre breasts panted together, the violently dilated nostrils quivered, the eyes stared stonily up-hill. They passed me within six inches, without a glance, with that complete, deathlike indifference of unhappy savages. Behind this raw matter one of the reclaimed, the product of the new forces at work, strolled despondently, carrying a rifle by its middle. He had a uniform jacket with one button off, and seeing a white man on the path, hoisted his weapon to his shoulder with alacrity. This was simple prudence, white men being so much alike at a distance that he could not tell who I might be. He was speedily reassured, and with a large, white, rascally grin, and a glance at his charge, seemed to take me into partnership in his exalted trust. After all, I also was a part of the great cause of these high and just proceedings.

"Instead of going up, I turned and descended to the left. My idea was to let that chain-gang get out of sight before I climbed the hill. You know I am not particularly tender; I've had to strike and to fend off. I've had to resist and to attack sometimes—that's only one way of resisting—without counting the exact cost, according to the demands of such sort of life as

I had blundered into. I've seen the devil of violence, and the devil of greed, and the devil of hot desire; but, by all the stars! these were strong, lusty, red-eyed devils, that swayed and drove men—men, I tell you. But as I stood on this hillside, I foresaw that in the blinding sunshine of that land I would become acquainted with a flabby, pretending, weak-eyed devil of a rapacious and pitiless folly. How insidious he could be, too, I was only to find out several months later and a thousand miles farther. For a moment I stood appalled, as though by a warning. Finally I descended the hill, obliquely, towards the trees I had seen.

"I avoided a vast artificial hole somebody had been digging on the slope, the purpose of which I found it impossible to divine. It wasn't a quarry or a sandpit, anyhow. It was just a hole. It might have been connected with the philanthropic desire of giving the criminals something to do. I don't know. Then I nearly fell into a very narrow ravine, almost no more than a scar in the hillside. I discovered that a lot of imported drainage-pipes for the settlement had been tumbled in there. There wasn't one that was not broken. It was a wanton smash-up. At last I got under the trees. My purpose was to stroll into the shade for a moment; but no sooner within than it seemed to me I had stepped into the gloomy circle of some Inferno. The rapids were near, and an uninterrupted, uniform, headlong, rushing noise filled the mournful stillness of the grove, where not a breath stirred, not a leaf moved, with a mysterious sound—as though the tearing pace of the launched earth had suddenly become audible.

"Black shapes crouched, lay, sat between the trees leaning against the trunks, clinging to the earth, half coming out, half effaced within the dim light, in all the attitudes of pain, abandonment, and despair. Another mine on the cliff went off, followed by a slight shudder of the soil under my feet. The work was going on. The work! And this was the place where some of the helpers had withdrawn to die.

"They were dying slowly—it was very clear. They were not enemies, they were not criminals, they were nothing earthly now—nothing but black shadows of disease and starvation, lying confusedly in the greenish gloom. Brought from all the recesses of the coast in all the legality of time contracts, lost in uncongenial surroundings, fed on unfamiliar food, they sickened, became inefficient, and were then allowed to crawl away and rest. These moribund shapes were free as air—and nearly as thin. I began to distinguish the gleam of the eyes under the trees. Then, glancing down, I saw a face near my hand. The black bones reclined at full length with one shoulder against the tree, and slowly the eyelids rose and the sunken eyes looked up at me, enormous and vacant, a kind of blind, white flicker in the depths of the orbs, which died out slowly. The man seemed young—almost a boy—but you know with them it's hard to tell. I found nothing else to do but to offer him one of my good Swede's ship's biscuits I had in my pocket. The fingers closed slowly on it and held—there was no other movement and no other glance. He had tied a bit of white worsted round his neck—Why? Where did he get it? Was it a badge—an ornament—a charm—a propitiatory act? Was there any idea at all connected with it? It looked startling round his black neck, this bit of white thread from beyond the seas.

"Near the same tree two more bundles of acute angles sat with their legs drawn up. One, with his chin propped on his knees, stared at nothing, in an intolerable and appalling manner: his brother phantom rested its forehead, as if overcome with a great weariness; and all about others were scattered in every pose of contorted collapse, as in some picture of a massacre or a pestilence. While I stood horror-struck, one of these creatures rose to his hands and knees, and went off on all-fours towards the river to drink. He lapped out of his hand, then sat up in the sunlight, crossing his shins in front of him, and after a time let his woolly head fall on his breastbone.

"I didn't want any more loitering in the shade, and I made haste towards the station. When near the buildings I met a white man, in such an unexpected elegance of get-up that in the first moment I took him for a sort of vision. I saw a high starched collar, white cuffs, a light alpaca jacket, snowy trousers, a clear necktie, and varnished boots. No hat. Hair parted, brushed, oiled, under a green-lined parasol held in a big white hand. He was amazing, and had a penholder behind his ear.

"I shook hands with this miracle, and I learned he was the Company's chief accountant, and that all the bookkeeping was done at this station. He had come out for a moment, he said, 'to get a breath of fresh air.' The expression sounded wonderfully odd, with its suggestion of sedentary desk-life. I wouldn't have mentioned the fellow to you at all, only it was from his lips that I first heard the name of the man who is so indissolubly connected with the memories of that time. Moreover, I respected the fellow. Yes; I respected his collars, his vast cuffs, his brushed hair. His appearance was certainly that of a hairdresser's dummy; but in the great demoralization of the land he kept up his appearance. That's backbone. His starched collars and got-up shirt-fronts were achievements of character. He had been out nearly three years; and later, I could not help asking him how he managed to sport such linen. He had just the faintest blush, and said modestly, 'I've been teaching one of the native women about the station. It was difficult. She had a distaste for the work.' Thus this man had verily accomplished something. And he was devoted to his books, which were in apple-pie order.

"Everything else in the station was in a muddle—heads, things, buildings. Strings of dusty niggers with splay feet arrived and departed; a stream of manufactured goods, rubbishy cottons, beads, and brass-wire sent into the depths of darkness, and in return came a precious trickle of ivory.

"I had to wait in the station for ten days—an eternity. I lived in a hut in the yard, but to be out of the chaos I would sometimes get into the accountant's office. It was built of horizontal planks, and so badly put together that, as he bent over his high desk, he was barred from neck to heels with narrow strips of sunlight. There was no need to open the big shutters to see. It was hot there, too; big flies buzzed fiendishly, and did not sting, but stabbed. I sat generally on the floor, while, of faultless appearance (and even slightly scented), perching on a high stool, he wrote. Sometimes he stood up for exercise. When a truckle-bed with a sick man (some invalid agent from up-country) was put in there, he exhibited a gentle annoyance. 'The groans of this sick person,' he said, 'distract my attention. And without that it is extremely difficult to guard against clerical errors in this climate.'

"One day he remarked, without lifting his head, 'In the interior you will no doubt meet Mr. Kurtz.' On my asking who Mr. Kurtz was, he said he was a first-class agent; and seeing my disappointment at this information, he added slowly, laying down his pen, 'He is a very remarkable person.' Further questions elicited from him that Mr. Kurtz was at present in charge of a trading post, a very important one, in the true ivory-country, at 'the very bottom of there. Sends in as much ivory as all the others put together . . .' He began to write again. The sick man was too ill to groan. The flies buzzed in a great peace.

"Suddenly there was a growing murmur of voices and a great tramping of feet. A caravan had come in. A violent babble of uncouth sounds burst out on the other side of the planks. All the carriers were speaking together, and in the midst of the uproar the lamentable voice of the chief agent was heard 'giving it up' tearfully for the twentieth time that day . . . He rose slowly. 'What a frightful row,' he said. He crossed the room gently to look at the sick man, and returning, said to me, 'He does not hear.' 'What! Dead?' I asked, startled. 'No, not yet,' he answered, with great composure. Then, alluding with a toss of the head to the tumult in the station-yard, 'When one has got to make correct entries, one comes to hate those savages—

hate them to the death.' He remained thoughtful for a moment. 'When you see Mr. Kurtz,' he went on, 'tell him from me that everything here'—he glanced at the desk—'is very satisfactory. I don't like to write to him—with those messengers of ours you never know who may get hold of your letter—at that Central Station.' He stared at me for a moment with his mild, bulging eyes. 'Oh, he will go far, very far,' he began again. 'He will be a somebody in the Administration before long. They, above—the Council in Europe, you know—mean him to be.'

"He turned to his work. The noise outside had ceased, and presently in going out I stopped at the door. In the steady buzz of flies the homeward-bound agent was lying flushed and insensible; the other, bent over his books, was making correct entries of perfectly correct transactions; and fifty feet below the doorstep I could see the still tree-tops of the grove of death.

"Next day I left that station at last, with a caravan of sixty men, for a two-hundred-mile tramp.

"No use telling you much about that. Paths, paths, everywhere; a stamped-in network of paths spreading over the empty land, through long grass, through burnt grass, through thickets, down and up chilly ravines, up and down stony hills ablaze with heat; and a solitude, a solitude, nobody, not a hut. The population had cleared out a long time ago. Well, if a lot of mysterious niggers armed with all kinds of fearful weapons suddenly took to travelling on the road between Deal and Gravesend, catching the yokels right and left to carry heavy loads for them, I fancy every farm and cottage thereabouts would get empty very soon. Only here the dwellings were gone, too. Still I passed through several abandoned villages. There's something pathetically childish in the ruins of grass walls. Day after day, with the stamp and shuffle of sixty pair of bare feet behind me, each pair under a 60-lb. load. Camp, cook, sleep, strike camp, march. Now and then a carrier dead in harness, at rest in the long grass near the path, with an empty water-gourd and his long staff lying by his side. A great silence around and above. Perhaps on some quiet night the tremor of far-off drums, sinking, swelling, a tremor vast, faint; a sound weird, appealing, suggestive, and wild—and perhaps with as profound a meaning as the sound of bells in a Christian country. Once a white man in an unbuttoned uniform, camping on the path with an armed escort of lank Zanzibaris, very hospitable and festive—not to say drunk. Was looking after the upkeep of the road, he declared. Can't say I saw any road or any upkeep, unless the body of a middle-aged negro, with a bullet-hole in the forehead, upon which I absolutely stumbled three miles farther on, may be considered as a permanent improvement. I had a white companion, too, not a bad chap, but rather too fleshy and with the exasperating habit of fainting on the hot hillsides, miles away from the least bit of shade and water. Annoying, you know, to hold your own coat like a parasol over a man's head while he is coming-to. I couldn't help asking him once what he meant by coming there at all. 'To make money, of course. What do you think?' he said, scornfully. Then he got fever, and had to be carried in a hammock slung under a pole. As he weighed sixteen stone I had no end of rows with the carriers. They jibbed, ran away, sneaked off with their loads in the night—quite a mutiny. So, one evening, I made a speech in English with gestures, not one of which was lost to the sixty pairs of eyes before me, and the next morning I started the hammock off in front all right. An hour afterwards I came upon the whole concern wrecked in a bush—man, hammock, groans, blankets, horrors. The heavy pole had skinned his poor nose. He was very anxious for me to kill somebody, but there wasn't the shadow of a carrier near. I remembered the old doctor—'It would be interesting for science to watch the mental changes of individuals, on the spot.' I felt I was becoming scientifically interesting. However, all that is to no purpose. On the fifteenth day I came in sight of the big river again, and hobbled into the Central

Station. It was on a back water surrounded by scrub and forest, with a pretty border of smelly mud on one side, and on the three others enclosed by a crazy fence of rushes. A neglected gap was all the gate it had, and the first glance at the place was enough to let you see the flabby devil was running that show. White men with long staves in their hands appeared languidly from amongst the buildings, strolling up to take a look at me, and then retired out of sight somewhere. One of them, a stout, excitable chap with black moustaches, informed me with great volubility and many digressions, as soon as I told him who I was, that my steamer was at the bottom of the river. I was thunderstruck. What, how, why? Oh, it was 'all right.' The 'manager himself' was there. All quite correct. 'Everybody had behaved splendidly! splendidly!'—'You must,' he said in agitation, 'go and see the general manager at once. He is waiting!'

"I did not see the real significance of that wreck at once. I fancy I see it now, but I am not sure—not at all. Certainly the affair was too stupid—when I think of it—to be altogether natural. Still . . . But at the moment it presented itself simply as a confounded nuisance. The steamer was sunk. They had started two days before in a sudden hurry up the river with the manager on board, in charge of some volunteer skipper, and before they had been out three hours they tore the bottom out of her on stones, and she sank near the south bank. I asked myself what I was to do there, now my boat was lost. As a matter of fact, I had plenty to do in fishing my command out of the river. I had to set about it the very next day. That, and the repairs when I brought the pieces to the station, took some months.

"My first interview with the manager was curious. He did not ask me to sit down after my twenty-mile walk that morning. He was commonplace in complexion, in feature, in manners, and in voice. He was of middle size and of ordinary build. His eyes, of the usual blue, were perhaps remarkably cold, and he certainly could make his glance fall on one as trenchant and heavy as an axe. But even at these times the rest of his person seemed to disclaim the intention. Otherwise there was only an indefinable, faint expression of his lips, something stealthy—a smile—not a smile—I remember it, but I can't explain. It was unconscious, this smile was, though just after he had said something it got intensified for an instant. It came at the end of his speeches like a seal applied on the words to make the meaning of the commonest phrase appear absolutely inscrutable. He was a common trader, from his youth up employed in these parts —nothing more. He was obeyed, yet he inspired neither love nor fear, nor even respect. He inspired uneasiness. That was it! Uneasiness. Not a definite mistrust—just uneasiness—nothing more. You have no idea how effective such a . . . a . . . faculty can be. He had no genius for organizing, for initiative, or for order even. That was evident in such things as the deplorable state of the station. He had no learning, and no intelligence. His position had come to him—why? Perhaps because he was never ill . . . He had served three terms of three years out there . . . Because triumphant health in the general rout of constitutions is a kind of power in itself. When he went home on leave he rioted on a large scale—pompously. Jack ashore—with a difference—in externals only. This one could gather from his casual talk. He originated nothing, he could keep the routine going—that's all. But he was great. He was great by this little thing that it was impossible to tell what could control such a man. He never gave that secret away. Perhaps there was nothing within him. Such a suspicion made one pause—for out there there were no external checks. Once when various tropical diseases had laid low almost every 'agent' in the station, he was heard to say, 'Men who come out here should have no entrails.' He sealed the utterance with that smile of his, as though it had been a door opening into a darkness he had in his keeping. You fancied you had seen things—but the seal was on. When annoyed at meal-times by the constant quarrels of the white men about precedence, he ordered an immense round table to be made, for which a special house had to be built. This was the station's mess-room. Where he sat was the first place—the rest were nowhere. One felt this to be his unalterable conviction. He was neither civil nor uncivil. He was quiet. He allowed his 'boy'—an overfed young negro from the coast—to treat the white men, under his very eyes, with provoking insolence.

"He began to speak as soon as he saw me. I had been very long on the road. He could not wait. Had to start without me. The up-river stations had to be relieved. There had been so many delays already that he did not know who was dead and who was alive, and how they got on—and so on, and so on. He paid no attention to my explanations, and, playing with a stick of sealing-wax, repeated several times that the situation was 'very grave, very grave.' There were rumours that a very important station was in jeopardy, and its chief, Mr. Kurtz, was ill. Hoped it was not true. Mr. Kurtz was . . . I felt weary and irritable. Hang Kurtz, I thought. I interrupted him by saying I had heard of Mr. Kurtz on the coast. 'Ah! So they talk of him down there,' he murmured to himself. Then he began again, assuring me Mr. Kurtz was the best agent he had, an exceptional man, of the greatest importance to the Company; therefore I could understand his anxiety. He was, he said, 'very, very uneasy.' Certainly he fidgeted on his chair a good deal, exclaimed, 'Ah, Mr. Kurtz!', broke the stick of sealing-wax and seemed dumfounded by the accident. Next thing he wanted to know 'how long it would take to' . . . I interrupted him again. Being hungry, you know, and kept on my feet, too, I was getting savage. 'How could I tell?' I said. 'I hadn't even seen the wreck yet—some months, no doubt.' All this talk seemed to me so futile. 'Some months,' he said. 'Well, let us say three months before we can make a start. Yes. That ought to do the affair.' I flung out of his hut (he lived all alone in a clay hut with a sort of verandah) muttering to myself my opinion of him. He was a chattering idiot. Afterwards I took it back when it was borne in upon me startlingly with what extreme nicety he had estimated the time requisite for the 'affair.'

"I went to work the next day, turning, so to speak, my back on that station. In that way only it seemed to me I could keep my hold on the redeeming facts of life. Still, one must look about sometimes; and then I saw this station, these men strolling aimlessly about in the sunshine of the yard. I asked myself sometimes what it all meant. They wandered here and there with their absurd long staves in their hands, like a lot of faithless pilgrims bewitched inside a rotten fence. The word 'ivory' rang in the air, was whispered, was sighed. You would think they were praying to it. A taint of imbecile rapacity blew through it all, like a whiff from some corpse. By Jove! I've never seen anything so unreal in my life. And outside, the silent wilderness surrounding this cleared speck on the earth struck me as something great and invincible, like evil or truth, waiting patiently for the passing away of this fantastic invasion.

"Oh, these months! Well, never mind. Various things happened. One evening a grass shed full of calico, cotton print, beads, and I don't know what else, burst into a blaze so suddenly that you would have thought the earth had opened to let an avenging fire consume all that trash. I was smoking my pipe quietly by my dismantled steamer, and saw them all cutting capers in the light, with their arms lifted high, when the stout man with moustaches came tearing down to the river, a tin pail in his hand, assured me that everybody was 'behaving splendidly, splendidly,' dipped about a quart of water and tore back again. I noticed there was a hole in the bottom of his pail.

"I strolled up. There was no hurry. You see the thing had gone off like a box of matches. It had been hopeless from the very first. The flame had leaped high, driven everybody back, lighted up everything—and collapsed. The shed was already a

heap of embers glowing fiercely. A nigger was being beaten near
by. They said he had caused the fire in some way; be that as
it may, he was screeching most horribly. I saw him, later, for
several days, sitting in a bit of shade looking very sick and trying
to recover himself: afterwards he arose and went out—and the
wilderness without a sound took him into its bosom again. As
I approached the glow from the dark I found myself at the back
of two men, talking. I heard the name of Kurtz pronounced,
then the words, 'take advantage of this unfortunate accident.'
One of the men was the manager. I wished him a good evening.
'Did you ever see anything like it—eh? It is incredible,' he said,
and walked off. The other man remained. He was a first-class
agent, young, gentlemanly, a bit reserved, with a forked little
beard and a hooked nose. He was stand-offish with the other
agents, and they on their side said he was the manager's spy
upon them. As to me, I had hardly ever spoken to him before.
We got into talk, and by-and-by we strolled away from the
hissing ruins. Then he asked me to his room, which was in the
main building of the station. He struck a match, and I perceived
that this young aristocrat had not only a silver-mounted dressing-
case but also a whole candle all to himself. Just at that time
the manager was the only man supposed to have any right to
candles. Native mats covered the clay walls; a collection of spears,
assegais, shields, knives was hung up in trophies. The business
intrusted to this fellow was the making of bricks—so I had
been informed; but there wasn't a fragment of a brick anywhere
in the station, and he had been there more than a year—waiting.
It seems he could not make bricks without something, I don't
know what—straw maybe. Anyways, it could not be found there,
and as it was not likely to be sent from Europe, it did not appear
clear to me what he was waiting for. An act of special creation
perhaps. However, they were all waiting—all the sixteen or
twenty pilgrims of them—for something; and upon my word
it did not seem an uncongenial occupation, from the way they

took it, though the only thing that ever came to them was disease—as far as I could see. They beguiled the time by back-biting and intriguing against each other in a foolish kind of way. There was an air of plotting about that station, but nothing came of it, of course. It was as unreal as everything else—as the philanthropic pretence of the whole concern, as their talk, as their government, as their show of work. The only real feeling was a desire to get appointed to a trading-post where ivory was to be had, so that they could earn percentages. They intrigued and slandered and hated each other only on that account—but as to effectually lifting a little finger—oh, no. By heavens! There is something after all in the world allowing one man to steal a horse while another must not look at a halter. Steal a horse straight out. Very well. He has done it. Perhaps he can ride. But there is a way of looking at a halter that would provoke the most charitable of saints into a kick.

"I had no idea why he wanted to be sociable, but as we chatted in there it suddenly occurred to me the fellow was trying to get at something—in fact, pumping me. He alluded constantly to Europe, to the people I was supposed to know there—putting leading questions as to my acquaintances in the sepulchral city, and so on. His little eyes glittered like mica discs—with curiosity—though he tried to keep up a bit of superciliousness. At first I was astonished, but very soon I became awfully curious to see what he would find out from me. I couldn't possibly imagine what I had in me to make it worth his while. It was very pretty to see how he baffled himself, for in truth my body was full only of chills, and my head had nothing in it but that wretched steamboat business. It was evident he took me for a perfectly shameless prevaricator. At last he got angry, and to conceal a movement of furious annoyance, he yawned. I rose. Then I noticed a small sketch in oils, on a panel, representing a woman, draped and blindfolded, carrying a lighted torch. The background was sombre—almost black. The movement of the woman was stately, and the effect of the torch-light on the face was sinister.

"It arrested me, and he stood by civilly, holding an empty half-pint champagne bottle (medical comforts) with the candle stuck in it. To my question he said Mr. Kurtz had painted this—in this very station more than a year ago—while waiting for means to go to his trading-post. 'Tell me, pray,' said I, 'who is this Mr. Kurtz?'

" 'The chief of the Inner Station,' he answered in a short tone, looking away. 'Much obliged,' I said, laughing. 'And you are the brickmaker of the Central Station. Everyone knows that.' He was silent for a while. 'He is a prodigy,' he said at last. 'He is an emissary of pity, and science, and progress, and devil knows what else. We want,' he began to declaim suddenly, 'for the guidance of the cause intrusted to us by Europe, so to speak, higher intelligence, wide sympathies, a singleness of purpose.' 'Who says that?' I asked. 'Lots of them,' he replied. 'Some even write that; and so *he* comes here, a special being, as you ought to know.' 'Why ought I to know?' I interrupted, really surprised. He paid no attention. 'Yes. Today he is chief of the best station, next year he will be assistant-manager, two years more and . . . but I daresay you know what he will be in two years' time. You are of the new gang—the gang of virtue. The same people who sent him specially also recommended you. Oh, don't say no. I've my own eyes to trust.' Light dawned upon me. My dear aunt's influential acquaintances were producing an unexpected effect upon that young man. I nearly burst into a laugh. 'Do you read the Company's confidential correspondence?' I asked. He hadn't a word to say. It was great fun. 'When Mr. Kurtz,' I continued severely, 'is General Manager, you won't have the opportunity.'

"He blew the candle out suddenly, and we went outside. The moon had risen. Black figures strolled about listlessly, pouring water on the glow, whence proceeded a sound of hissing; steam ascended in the moonlight, the beaten nigger groaned some-where. 'What a row the brute makes!' said the indefatigable man with the moustaches, appearing near us. 'Serve him right. Transgression—punishment—bang! Pitiless, pitiless. That's the only way. This will prevent all conflagrations for the future. I was just telling the manager . . .' He noticed my companion, and became crestfallen all at once. 'Not in bed yet,' he said, with a kind of servile heartiness; 'it's so natural. Ha! Danger—agitation.' He vanished. I went on to the river side, and the other followed me. I heard a scathing murmur at my ear, 'Heap of muffs—go to.' The pilgrims could be seen in knots gestic-ulating, discussing. Several had still their staves in their hands. I verily believe they took these sticks to bed with them. Beyond the fence the forest stood up spectrally in the moonlight, and through the dim stir, through the faint sounds of that lamentable courtyard, the silence of the land went home to one's very heart—its mystery, its greatness, the amazing reality of its con-cealed life. The hurt nigger moaned feebly somewhere near by, and then fetched a deep sigh that made me mend my pace away from there. I felt a hand introducing itself under my arm. 'My dear sir,' said the fellow, 'I don't want to be misunderstood, and especially by you, who will see Mr. Kurtz long before I can have that pleasure. I wouldn't like him to get a false idea of my disposition . . .'

"I let him run on, this papier-maché Mephistopheles, and it seemed to me that if I tried I could poke my forefinger through him, and would find nothing inside but a little loose dirt, maybe. He, don't you see, had been planning to be assistant-manager by-and-by under the present man, and I could see that the coming of that Kurtz had upset them both not a little. He talked precipitately, and I did not try to stop him. I had my shoulders against the wreck of my steamer, hauled up on the slope like a carcass of some big river animal. The smell of mud, of primeval mud, by Jove! was in my nostrils, the high stillness

of primeval forest was before my eyes; there were shiny patches
on the black creek. The moon had spread over everything a thin
layer of silver—over the rank grass, over the mud, upon the
wall of matted vegetation standing higher than the wall of a
temple, over the great river I could see through a sombre gap
glittering, glittering, as it flowed broadly by without a murmur.
All this was great, expectant, mute, while the man jabbered
about himself. I wondered whether the stillness on the face of
the immensity looking at us two were meant as an appeal or
as a menace. What were we who had strayed in here? Could
we handle that dumb thing, or would it handle us? I felt how
big, how confoundedly big, was that thing that couldn't talk,
and perhaps was deaf as well. What was in there? I could see
a little ivory coming out from there, and I had heard Mr. Kurtz
was in there. I had heard enough about it, too—God knows!
Yet somehow it didn't bring any image with it—no more than
if I had been told an angel or a fiend was in there. I believed
it in the same way one of you might believe there are inhabitants
in the planet Mars. I knew once a Scotch sailmaker who was
certain, dead sure, there were people in Mars. If you asked him
for some idea how they looked and behaved, he would get shy
and mutter something about 'walking on all fours.' If you as
much as smiled, he would—though a man of sixty—offer to
fight you. I would not have gone so far as to fight for Kurtz,
but I went for him near enough to a lie. You know I hate,
detest, and can't bear a lie, not because I am straighter than
the rest of us, but simply because it appals me. There is a taint
of death, a flavour of mortality in lies—which is exactly what
I hate and detest in the world—what I want to forget. It makes
me miserable and sick, like biting something rotten would do.
Temperament, I suppose. Well, I went near enough to it by
letting the young fool there believe anything he liked to imagine
as to my influence in Europe. I became in an instant as much
of a pretence as the rest of the bewitched pilgrims. This simply

because I had a notion it somehow would be of help to that Kurtz whom at the time I did not see—you understand. He was just a word for me. I did not see the man in the name any more than you do. Do you see him? Do you see the story? Do you see anything? It seems to me I am trying to tell you a dream—making a vain attempt, because no relation of a dream can convey the dream-sensation, that commingling of absurdity, surprise, and bewilderment in a tremor of struggling revolt, that notion of being captured by the incredible which is of the very essence of dreams . . ."

He was silent for a while.

". . . No, it is impossible; it is impossible to convey the life-sensation of any given epoch of one's existence—that which makes its truth, its meaning—its subtle and penetrating essence. It is impossible. We live, as we dream—alone . . ."

He paused again as if reflecting, then added—

"Of course in this you fellows see more than I could then. You see me, whom you know . . ."

It had become so pitch dark that we listeners could hardly see one another. For a long time already he, sitting apart, had been no more to us than a voice. There was not a word from anybody. The others might have been asleep, but I was awake. I listened, I listened on the watch for the sentence, for the word, that would give me the clue to the faint uneasiness inspired by this narrative that seemed to shape itself without human lips in the heavy night-air of the river.

". . . Yes—I let him run on," Marlow began again, "and think what he pleased about the powers that were behind me. I did! And there was nothing behind me! There was nothing but that wretched, old, mangled steamboat I was leaning against, while he talked fluently about 'the necessity for every man to get on.' 'And when one comes out here, you conceive, it is not to gaze at the moon.' Mr. Kurtz was a 'universal genius,' but even a genius would find it easier to work with 'adequate tools—intelligent men.' He did not make bricks—why, there was a physical impossibility in the way—as I was well aware; and if he did secretarial work for the manager, it was because 'no sensible man rejects wantonly the confidence of his superiors.' Did I see it? I saw it. What more did I want? What I really wanted was rivets, by heaven! Rivets. To get on with the work—to stop the hole. Rivets I wanted. There were cases of them down at the coast—cases—piled up—burst—split! You kicked a loose rivet at every second step in that station yard on the hillside. Rivets had rolled into the grove of death. You could fill your pockets with rivets for the trouble of stooping down—and there wasn't one rivet to be found where it was wanted. We had plates that would do, but nothing to fasten them with. And every week the messenger, a lone negro, letter-bag on shoulder and staff in hand, left our station for the coast. And several times a week a coast caravan came in with trade goods—ghastly glazed calico that made you shudder only to look at it, glass beads, value about a penny a quart, confounded spotted cotton handkerchiefs. And no rivets. Three carriers could have brought all that was wanted to set that steamboat afloat.

"He was becoming confidential now, but I fancy my unresponsive attitude must have exasperated him at last, for he judged it necessary to inform me he feared neither God nor devil, let alone any mere man. I said I could see that very well, but what I wanted was a certain quantity of rivets—and rivets were what really Mr. Kurtz wanted, if he had only known it. Now letters went to the coast every week . . . 'My dear sir,' he cried, 'I write from dictation.' I demanded rivets. There was a way—for an intelligent man. He changed his manner; became very cold, and suddenly began to talk about a hippopotamus; wondered whether sleeping on board the steamer (I stuck to my salvage night and day) I wasn't disturbed. There was an old hippo that had the bad habit of getting out on the bank and roaming at night over the station grounds. The pilgrims used to turn out in a body and empty every rifle they could lay hands on at him. Some even had sat up o' nights for him. All this energy was wasted, though. 'That animal has a charmed life,' he said; 'but you can say this only of brutes in this country. No man—you apprehend me?—no man here bears a charmed life.' He stood there for a moment in the moonlight with his delicate hooked nose set a little askew, and his mica eyes glittering without a wink, then, with a curt Good-night, he strode off. I could see he was disturbed and considerably puzzled, which made me feel more hopeful than I had been for days. It was a great comfort to turn from that chap to my influential friend, the battered, twisted, ruined, tin-pot steamboat. I clambered on board. She rang under my feet like an empty Huntley & Palmers biscuit-tin kicked along a gutter; she was nothing so solid in make, and rather less pretty in shape, but I had expended enough hard work on her to make me love her. No influential friend would have served me better. She had given me a chance to come out a bit—to find out what I could do. No, I don't like work. I had rather laze about and think of all the fine things that can be done. I don't like work, no man does—but I like what is in the work, the chance to find yourself. Your own reality—for yourself, not for others—what no other man can ever know. They can only see the mere show, and never can tell what it really means.

"I was not surprised to see somebody sitting aft, on the deck, with his legs dangling over the mud. You see I rather chummed with the few mechanics there were in that station, whom the other pilgrims naturally despised—on account of their imperfect manners, I suppose. This was the foreman—a boiler-maker by trade—a good worker. He was a lank, bony, yellow-faced man, with big intense eyes. His aspect was worried, and his head was as bald as the palm of my hand; but his hair in falling seemed to have stuck to his chin, and had prospered in the new locality, for his beard hung down to his waist. He was a widower with

six young children (he had left them in charge of a sister of his
to come out there), and the passion of his life was pigeon-flying.
He was an enthusiast and a connoisseur. He would rave about
pigeons. After work hours he used sometimes to come over from
his hut for a talk about his children and his pigeons; at work,
when he had to crawl in the mud under the bottom of the
steamboat, he would tie up that beard of his in a kind of white
serviette he brought for the purpose. It had loops to go over
his ears. In the evening he could be seen squatted on the bank
rinsing that wrapper in the creek with great care, then spreading
it solemnly on a bush to dry.

"I slapped him on the back and shouted 'We shall have
rivets!' He scrambled to his feet exclaiming 'No! Rivets!' as
though he couldn't believe his ears. Then in a low voice, 'You
. . . eh?' I don't know why we behaved like lunatics. I put my
finger to the side of my nose and nodded mysteriously. 'Good
for you!' he cried, snapped his fingers above his head, lifting
one foot. I tried a jig. We capered on the iron deck. A frightful
clatter came out of that hulk, and the virgin forest on the other
bank of the creek sent it back in a thundering roll upon the
sleeping station. It must have made some of the pilgrims sit up
in their hovels. A dark figure obscured the lighted doorway of
the manager's hut, vanished, then, a second or so after, the
doorway itself vanished, too. We stopped, and the silence driven
away by the stamping of our feet flowed back again from the
recesses of the land. The great wall of vegetation, an exuberant
and entangled mass of trunks, branches, leaves, boughs, festoons,
motionless in the moonlight, was like a rioting invasion of
soundless life, a rolling wave of plants, piled up, crested, ready
to topple over the creek, to sweep every little man of us out of
his little existence. And it moved not. A deadened burst of
mighty splashes and snorts reached us from afar, as though an
ichthyosaurus had been taking a bath of glitter in the great river.
'After all,' said the boiler-maker in a reasonable tone, 'why

shouldn't we get the rivets?' Why not, indeed! I did not know of any reason why we shouldn't. 'They'll come in three weeks,' I said, confidently.

"But they didn't. Instead of rivets there came an invasion, an infliction, a visitation. It came in sections during the next three weeks, each section headed by a donkey carrying a white man in new clothes and tan shoes, bowing from that elevation right and left to the impressed pilgrims. A quarrelsome band of footsore sulky niggers trod on the heels of the donkey; a lot of tents, campstools, tin boxes, white cases, brown bales would be shot down in the courtyard, and the air of mystery would deepen a little over the muddle of the station. Five such instalments came, with their absurd air of disorderly flight with the loot of innumerable outfit shops and provision stores, that, one would think, they were lugging, after a raid, into the wilderness for equitable division. It was an inextricable mess of things decent in themselves but that human folly made look like spoils of thieving.

"This devoted band called itself the Eldorado Exploring Expedition, and I believe they were sworn to secrecy. Their talk, however, was the talk of sordid buccaneers: it was reckless without hardihood, greedy without audacity, and cruel without courage; there was not an atom of foresight or of serious intention in the whole batch of them, and they did not seem aware these things are wanted for the work of the world. To tear treasure out of the bowels of the land was their desire, with no more moral purpose at the back of it than there is in burglars breaking into a safe. Who paid the expenses of the noble enterprise I don't know; but the uncle of our manager was leader of that lot.

"In exterior he resembled a butcher in a poor neighbourhood, and his eyes had a look of sleepy cunning. He carried his fat paunch with ostentation on his short legs, and during the time his gang infested the station spoke to no one but his nephew.

You could see these two roaming about all day long with their heads close together in an everlasting confab.

"I had given up worrying myself about the rivets. One's capacity for that kind of folly is more limited than you would suppose. I said Hang!—and let things slide. I had plenty of time for meditation, and now and then I would give some thought to Kurtz. I wasn't very interested in him. No. Still, I was curious to see whether this man, who had come out equipped with moral ideas of some sort, would climb to the top after all and how he would set about his work when there."

2

"One evening as I was lying flat on the deck of my steamboat, I heard voices approaching—and there were the nephew and the uncle strolling along the bank. I laid my head on my arm again, and had nearly lost myself in a doze, when somebody said in my ear, as it were: 'I am as harmless as a little child, but I don't like to be dictated to. Am I the manager—or am I not? I was ordered to send him there. It's incredible.' . . . I became aware that the two were standing on the shore alongside the forepart of the steamboat, just below my head. I did not move; it did not occur to me to move: I was sleepy. 'It *is* unpleasant,' grunted the uncle. 'He has asked the Administration to be sent there,' said the other, 'with the idea of showing what he could do; and I was instructed accordingly. Look at the influence that man must have. Is it not frightful?' They both agreed it was frightful, then made several bizarre remarks: 'Make rain and fine weather—one man—the Council—by the nose'—bits of absurd sentences that got the better of my drowsiness, so that I had pretty near the whole of my wits about me when the uncle said, 'The climate may do away with this difficulty for you. Is he alone there?' 'Yes,' answered the manager; 'he sent his assistant down the river with a note to me in these terms: "Clear this poor devil out of the country, and don't bother sending more of that sort. I had rather be alone than have the kind of men you can dispose of with me." It was more than a year ago. Can you imagine such impudence!' 'Anything since then?' asked the other, hoarsely. 'Ivory,' jerked the nephew; 'lots of it—prime sort—lots—most annoying, from him.' 'And with that?' questioned the heavy rumble. 'Invoice,' was the reply fired out, so to speak. Then silence. They had been talking about Kurtz.

"I was broad awake by this time, but, lying perfectly at ease, remained still, having no inducement to change my position. 'How did that ivory come all this way?' growled the elder man, who seemed very vexed. The other explained that it had come with a fleet of canoes in charge of an English half-caste clerk Kurtz had with him; that Kurtz had apparently intended to return himself, the station being by that time bare of goods and stores, but after coming three hundred miles, had suddenly decided to go back, which he started to do alone in a small dugout with four paddlers, leaving the half-caste to continue down the river with the ivory. The two fellows there seemed astounded at anybody attempting such a thing. They were at a loss for an adequate motive. As to me, I seemed to see Kurtz for the first time. It was a distinct glimpse: the dugout, four paddling savages, and the lone white man turning his back suddenly on the headquarters, on relief, on thoughts of home —perhaps; setting his face towards the depths of the wilderness, towards his empty and desolate station. I did not know the motive. Perhaps he was just simply a fine fellow who stuck to his work for its own sake. His name, you understand, had not been pronounced once. He was 'that man.' The half-caste, who, as far as I could see, had conducted a difficult trip with great prudence and pluck, was invariably alluded to as 'that scoundrel.' The 'scoundrel' had reported that the 'man' had been very ill —had recovered imperfectly. . . . The two below me moved away then a few paces, and strolled back and forth at some little

distance. I heard: 'Military post—doctor—two hundred miles
—quite alone now—unavoidable delays—nine months—no
news—strange rumours.' They approached again, just as the
manager was saying, 'No one, as far as I know, unless a species
of wandering trader—a pestilential fellow, snapping ivory from
the natives.' Who was it they were talking about now? I gathered
in snatches that this was some man supposed to be in Kurtz's
district, and of whom the manager did not approve. 'We will
not be free from unfair competition till one of these fellows is
hanged for an example,' he said. 'Certainly,' grunted the other;
'get him hanged! Why not? Anything—anything can be done
in this country. That's what I say; nobody here, you understand,
here, can endanger your position. And why? You stand the
climate—you outlast them all. The danger is in Europe; but
there before I left I took care to—' They moved off and whispered, then their voices rose again. 'The extraordinary series of
delays is not my fault. I did my best.' The fat man sighed.
'Very sad.' 'And the pestiferous absurdity of his talk,' continued
the other; 'he bothered me enough when he was here. "Each
station should be like a beacon on the road towards better things,
a centre for trade of course, but also for humanizing, improving,
instructing." Conceive you—that ass! And he wants to be manager! No, it's—' Here he got choked by excessive indignation,
and I lifted my head the least bit. I was surprised to see how
near they were—right under me. I could have spat upon their
hats. They were looking on the ground, absorbed in thought.
The manager was switching his leg with a slender twig: his
sagacious relative lifted his head. 'You have been well since you
came out this time?' he asked. The other gave a start. 'Who?
I? Oh! Like a charm—like a charm. But the rest—oh, my goodness! All sick. They die so quick, too, that I haven't the time
to send them out of the country—it's incredible!' 'H'm. Just
so,' grunted the uncle. 'Ah! my boy, trust to this—I say, trust
to this.' I saw him extend his short flipper of an arm for a

gesture that took in the forest, the creek, the mud, the river—seemed to beckon with a dishonouring flourish before the sunlit face of the land a treacherous appeal to the lurking death, to the hidden evil, to the profound darkness of its heart. It was so startling that I leaped to my feet and looked back at the edge of the forest, as though I had expected an answer of some sort to that black display of confidence. You know the foolish notions that come to one sometimes. The high stillness confronted these two figures with its ominous patience, waiting for the passing away of a fantastic invasion.

"They swore aloud together—out of sheer fright, I believe—then pretending not to know anything of my existence, turned back to the station. The sun was low; and leaning forward side by side, they seemed to be tugging painfully uphill their two ridiculous shadows of unequal length, that trailed behind them slowly over the tall grass without bending a single blade.

"In a few days the Eldorado Expedition went into the patient wilderness, that closed upon it as the sea closes over a diver. Long afterwards the news came that all the donkeys were dead. I know nothing as to the fate of the less valuable animals. They, no doubt, like the rest of us, found what they deserved. I did not inquire. I was then rather excited at the prospect of meeting Kurtz very soon. When I say very soon I mean it comparatively. It was just two months from the day we left the creek when we came to the bank below Kurtz's station.

"Going up that river was like travelling back to the earliest beginnings of the world, when vegetation rioted on the earth and the big trees were kings. An empty stream, a great silence, an impenetrable forest. The air was warm, thick, heavy, sluggish. There was no joy in the brilliance of sunshine. The long stretches of the waterway ran on, deserted, into the gloom of overshadowed distances. On silvery sandbanks hippos and alligators sunned themselves side by side. The broadening waters flowed through a mob of wooded islands; you lost your way on that river as you would in a desert, and butted all day long against shoals, trying to find the channel, till you thought yourself bewitched and cut off for ever from everything you had known once—somewhere—far away—in another existence perhaps. There were moments when one's past came back to one, as it will sometimes when you have not a moment to spare to yourself; but it came in the shape of an unrestful and noisy dream, remembered with wonder amongst the overwhelming realities of this strange world of plants, and water, and silence. And this stillness of life did not in the least resemble a peace. It was the stillness of an implacable force brooding over an inscrutable intention. It looked at you with a vengeful aspect. I got used to it afterwards; I did not see it any more; I had no time. I had to keep guessing at the channel; I had to discern, mostly by inspiration, the signs of hidden banks; I watched for sunken stones; I was learning to clap my teeth smartly before my heart flew out, when I shaved by a fluke some infernal sly old snag that would have ripped the life out of the tin-pot steamboat and drowned all the pilgrims; I had to keep a look-out for the signs of dead wood we could cut up in the night for next day's steaming. When you have to attend to things of that sort, to the mere incidents of the surface, the reality—the reality, I tell you—fades. The inner truth is hidden—luckily, luckily. But I felt it all the same; I felt often its mysterious stillness watching me at my monkey tricks, just as it watches you fellows performing on your respective tight-ropes for—what is it? half-a-crown a tumble—"

"Try to be civil, Marlow," growled a voice, and I knew there was at least one listener awake besides myself.

"I beg your pardon. I forgot the heartache which makes up the rest of the price. And indeed what does the price matter, if the trick be well done? You do your tricks very well. And I didn't do badly either, since I managed not to sink that steamboat on my first trip. It's a wonder to me yet. Imagine a blindfolded man set to drive a van over a bad road. I sweated and shivered over that business considerably, I can tell you. After all, for a seaman, to scrape the bottom of the thing that's supposed to float all the time under his care is the unpardonable sin. No one may know of it, but you never forget the thump —eh? A blow on the very heart. You remember it, you dream of it, you wake up at night and think of it—years after—and go hot and cold all over. I don't pretend to say that steamboat floated all the time. More than once she had to wade for a bit, with twenty cannibals splashing around and pushing. We had enlisted some of these chaps on the way for a crew. Fine fellows—cannibals—in their place. They were men one could work with, and I am grateful to them. And, after all, they did not eat each other before my face: they had brought along a provision of hippo-meat which went rotten, and made the mystery of the wilderness stink in my nostrils. Phoo! I can sniff it now. I had the manager on board and three or four pilgrims with their staves—all complete. Sometimes we came upon a station close by the bank, clinging to the skirts of the unknown, and the white men rushing out of a tumbledown hovel, with great gestures of joy and surprise and welcome, seemed very strange—had the appearance of being held there captive by a spell. The word ivory would ring in the air for a while—and on we went again into the silence, along empty reaches, round the still bends, between the high walls of our winding way, reverberating in hollow claps the ponderous beat of the stern-wheel. Trees, trees, millions of trees, massive, immense, running up high; and at their foot, hugging the bank against the stream, crept the little begrimed steamboat, like a sluggish beetle crawling on the floor of a lofty portico. It made you feel very small, very lost, and yet it was not altogether depressing, that feeling. After all, if you were small, the grimy beetle crawled on—which was just what you wanted it to do. Where the pilgrims imagined it crawled to I don't know. To some place where they expected to get something, I bet! For me it crawled towards Kurtz—exclusively; but when the steam-pipes started leaking we crawled very slow. The reaches opened before us and closed behind, as if the forest had stepped leisurely across the water to bar the way for our return. We penetrated deeper and deeper into the heart of darkness. It was very quiet there. At night sometimes the roll of drums behind the curtain of trees would run up the river and remain sustained faintly, as if hovering in the air high over our heads, till the first break of day. Whether it meant war, peace, or prayer we could not tell. The dawns were heralded by the descent of a chill stillness; the wood-cutters slept, their fires burned low; the snapping of a twig would make you start. We were wanderers on prehistoric earth, on an earth that wore the aspect of an unknown planet. We could have fancied ourselves the first of men taking possession of an accursed inheritance, to be subdued at the cost of profound anguish and of excessive toil. But suddenly, as we struggled round a bend, there would be a glimpse of rush walls, of peaked grass-roofs, a burst of yells, a whirl of black limbs, a mass of hands clapping, of feet stamping, of bodies swaying, of eyes rolling, under the droop of heavy and motionless foliage. The steamer toiled along slowly on the edge of a black and incomprehensible frenzy. The prehistoric man was cursing us, praying to us, welcoming us—who could tell? We were cut off from the comprehension of our surroundings; we glided past like phantoms, wondering and secretly appalled, as sane men would be before an enthusiastic outbreak in a madhouse. We could not understand because we were too far and could not remember, because we were travelling in the night of first ages, of those ages that are gone, leaving hardly a sign—and no memories.

"The earth seemed unearthly. We are accustomed to look upon the shackled form of a conquered monster, but there—there you could look at a thing monstrous and free. It was unearthly, and the men were—No, they were not inhuman.

Well, you know, that was the worst of it—this suspicion of their not being inhuman. It would come slowly to one. They howled and leaped, and spun, and made horrid faces; but what thrilled you was just the thought of their humanity—like yours—the thought of your remote kinship with this wild and passionate uproar. Ugly. Yes, it was ugly enough; but if you were man enough you would admit to yourself that there was in you just the faintest trace of a response to the terrible frankness of that noise, a dim suspicion of there being a meaning in it which you—you so remote from the night of first ages—could comprehend. And why not? The mind of man is capable of anything—because everything is in it, all the past as well as all the future. What was there after all? Joy, fear, sorrow, devotion, valour, rage—who can tell?—but truth—truth stripped of its cloak of time. Let the fool gape and shudder—the man knows, and can look on without a wink. But he must at least be as much of a man as these on the shore. He must meet that truth with his own true stuff—with his own inborn strength. Principles won't do. Acquisitions, clothes, pretty rags—rags that would fly off at the first good shake. No; you want a deliberate belief. An appeal to me in this fiendish row—is there? Very well; I hear; I admit, but I have a voice, too, and for good or evil mine is the speech that cannot be silenced. Of course, a fool, what with sheer fright and fine sentiments, is always safe. Who's that grunting? You wonder I didn't go ashore for a howl and a dance? Well, no—I didn't. Fine sentiments, you say? Fine sentiments, be hanged! I had no time. I had to mess about with white-lead and strips of woollen blanket helping to put bandages on those leaky steam-pipes—I tell you. I had to watch the steering, and circumvent those snags, and get the tin-pot along by hook or by crook. There was surface-truth enough in these things to save a wiser man. And between whiles I had to look after the savage who was fireman. He was an improved specimen; he could fire up a vertical boiler. He was there below me, and, upon my word, to look at him was as edifying as seeing a dog in a parody of breeches and a feather hat, walking on his hind-legs. A few months of training had done for that really fine chap. He squinted at the steam-gauge and at the water-gauge with an evident effort of intrepidity—and he had filed teeth, too, the poor devil, and the wool of his pate shaved into queer patterns, and three ornamental scars on each of his cheeks. He ought to have been clapping his hands and stamping his feet on the bank, instead of which he was hard at work, a thrall to strange witchcraft, full of improving knowledge. He was useful because he had been instructed; and what he knew was this—that should the water in that transparent thing disappear, the evil spirit inside the boiler would get angry through the greatness of his thirst, and take a terrible vengeance. So he sweated and fired up and watched the glass fearfully (with an impromptu charm, made of rags, tied to his arm, and a piece of polished bone, as big as a watch, stuck flatways through his lower lip), while the wooded banks slipped past us slowly, the short noise was left behind, the interminable miles of silence—and we crept on, towards Kurtz. But the snags were thick, the water was treacherous and shallow, the boiler seemed indeed to have a sulky devil in it, and thus neither that fireman nor I had any time to peer into our creepy thoughts.

"Some fifty miles below the Inner Station we came upon a hut of reeds, an inclined and melancholy pole, with the unrecognizable tatters of what had been a flag of some sort flying from it, and a neatly stacked wood-pile. This was unexpected. We came to the bank, and on the stack of firewood found a flat piece of board with some faded pencil-writing on it. When deciphered it said: 'Wood for you. Hurry up. Approach cautiously.' There was a signature, but it was illegible—not Kurtz—a much longer word. Hurry up. Where? Up the river? 'Approach cautiously.' We had not done so. But the warning could not have been meant for the place where it could be only

found after approach. Something was wrong above. But what
—and how much? That was the question. We commented ad-
versely upon the imbecility of that telegraphic style. The bush
around said nothing, and would not let us look very far, either.
A torn curtain of red twill hung in the doorway of the hut, and
flapped sadly in our faces. The dwelling was dismantled; but
we could see a white man had lived there not very long ago.
There remained a rude table—a plank on two posts; a heap of
rubbish reposed in a dark corner, and by the door I picked up
a book. It had lost its covers, and the pages had been thumbed
into a state of extremely dirty softness; but the back had been
lovingly stitched afresh with white cotton thread, which looked
clean yet. It was an extraordinary find. Its title was, *An Inquiry
into some Points of Seamanship,* by a man, Tower, Towson—
some such name—Master in his Majesty's Navy. The matter
looked dreary reading enough, with illustrative diagrams and
repulsive tables of figures, and the copy was sixty years old. I
handled this amazing antiquity with the greatest possible ten-
derness, lest it should dissolve in my hands. Within, Towson
or Towser was inquiring earnestly into the breaking strain of
ships' chains and tackle, and other such matters. Not a very
enthralling book; but at the first glance you could see there a
singleness of intention, an honest concern for the right way of
going to work, which made these humble pages, thought out
so many years ago, luminous with another than a professional
light. The simple old sailor, with his talk of chains and purchases,
made me forget the jungle and the pilgrims in a delicious sensation of having come upon something unmistakably real. Such
a book being there was wonderful enough; but still more as-
tounding were the notes pencilled in the margin, and plainly
referring to the text. I couldn't believe my eyes! They were in
cipher! Yes, it looked like cipher. Fancy a man lugging with
him a book of that description into this nowhere and studying
it—and making notes—in cipher at that! It was an extravagant
mystery.

"I had been dimly aware for some time of a worrying noise, and when I lifted my eyes I saw the wood-pile was gone, and the manager, aided by all the pilgrims, was shouting at me from the river side. I slipped the book into my pocket. I assure you to leave off reading was like tearing myself away from the shelter of an old and solid friendship.

"I started the lame engine ahead. 'It must be this miserable trader—this intruder,' exclaimed the manager, looking back malevolently at the place we had left. 'He must be English,' I said. 'It will not save him from getting into trouble if he is not careful,' muttered the manager darkly. I observed with assumed innocence that no man was safe from trouble in this world.

"The current was more rapid now, the steamer seemed at her last gasp, the stern-wheel flopped languidly, and I caught myself listening on tiptoe for the next beat of the boat, for in sober truth I expected the wretched thing to give up every moment. It was like watching the last flickers of a life. But still we crawled. Sometimes I would pick out a tree a little way ahead to measure our progress towards Kurtz by, but I lost it invariably before we got abreast. To keep the eyes so long on one thing was too much for human patience. The manager displayed a beautiful resignation. I fretted and fumed and took to arguing with myself whether or no I would talk openly with Kurtz; but before I could come to any conclusion it occurred to me that my speech or my silence, indeed any action of mine, would be a mere futility. What did it matter what any one knew or ignored? What did it matter who was manager? One gets sometimes such a flash of insight. The essentials of this affair lay deep under the surface, beyond my reach, and beyond my power of meddling.

"Towards the evening of the second day we judged ourselves about eight miles from Kurtz's station. I wanted to push on; but the manager looked grave, and told me the navigation up there was so dangerous that it would be advisable, the sun being very low already, to wait where we were till next morning.

Moreover, he pointed out that if the warning to approach cautiously were to be followed, we must approach in daylight—not at dusk, or in the dark. This was sensible enough. Eight miles meant nearly three hours' steaming for us, and I could also see suspicious ripples at the upper end of the reach. Nevertheless, I was annoyed beyond expression at the delay, and most unreasonably, too, since one night more could not matter much after so many months. As we had plenty of wood, and caution was the word, I brought up in the middle of the stream. The reach was narrow, straight, with high sides like a railway cutting. The dusk came gliding into it long before the sun had set. The current ran smooth and swift, but a dumb immobility sat on the banks. The living trees, lashed together by the creepers and every living bush of the undergrowth, might have been changed into stone, even to the slenderest twig, to the lightest leaf. It was not sleep—it seemed unnatural, like a state of trance. Not the faintest sound of any kind could be heard. You looked on amazed, and began to suspect yourself of being deaf—then the night came suddenly, and struck you blind as well. About three in the morning some large fish leaped, and the loud splash made me jump as though a gun had been fired. When the sun rose there was a white fog, very warm and clammy, and more blinding than the night. It did not shift or drive; it was just there, standing all round you like something solid. At eight or nine, perhaps, it lifted as a shutter lifts. We had a glimpse of the towering multitude of trees, of the immense matted jungle, with the blazing little ball of the sun hanging over it—all perfectly still—and then the white shutter came down again, smoothly, as if sliding in greased grooves. I ordered the chain, which we had begun to heave in, to be paid out again. Before it stopped running with a muffled rattle, a cry, a very loud cry, as of infinite desolation, soared slowly in the opaque air. It ceased. A complaining clamour, modulated in savage discords, filled our ears. The sheer unexpectedness of it made my hair stir under my cap. I don't know how it struck the others: to me it seemed as though the mist itself had screamed, so suddenly, and apparently from all sides at once, did this tumultuous and mournful uproar arise. It culminated in a hurried outbreak of almost intolerably excessive shrieking, which stopped short, leaving us stiffened in a variety of silly attitudes, and obstinately listening to the nearly as appalling and excessive silence. 'Good God! What is the meaning—' stammered at my elbow one of the pilgrims, a little fat man, with sandy hair and red whiskers, who wore sidespring boots, and pink pyjamas tucked into his socks. Two others remained open-mouthed a whole minute, then dashed into the little cabin, to rush out incontinently and stand darting scared glances, with Winchesters at 'ready' in their hands. What we could see was just the steamer we were on, her outlines blurred as though she had been on the point of dissolving, and a misty strip of water, perhaps two feet broad, around her—and that was all. The rest of the world was nowhere, as far as our eyes and ears were concerned. Just nowhere. Gone, disappeared; swept off without leaving a whisper or a shadow behind.

"I went forward, and ordered the chain to be hauled in short, so as to be ready to trip the anchor and move the steamboat at once if necessary. 'Will they attack?' whispered an awed voice. 'We will be all butchered in this fog,' murmured another. The faces twitched with the strain, the hands trembled slightly, the eyes forgot to wink. It was very curious to see the contrast of expressions of the white men and of the black fellows of our crew, who were as much strangers to that part of the river as we, though their homes were only eight hundred miles away. The whites, of course greatly discomposed, had besides a curious look of being painfully shocked by such an outrageous row. The others had an alert, naturally interested expression; but their faces were essentially quiet, even those of the one or two who grinned as they hauled at the chain. Several exchanged short, grunting phrases, which seemed to settle the matter to their

satisfaction. Their headman, a young, broad-chested black, severely draped in dark-blue fringed cloths, with fierce nostrils
and his hair all done up artfully in oily ringlets, stood near me.
'Aha!' I said, just for good fellowship's sake. 'Catch 'im,' he
snapped, with a bloodshot widening of his eyes and a flash of
sharp teeth—'catch 'im. Give 'im to us.' 'To you, eh?' I asked;
'what would you do with them?' 'Eat 'im!' he said, curtly, and,
leaning his elbow on the rail, looked out into the fog in a
dignified and profoundly pensive attitude. I would no doubt
have been properly horrified, had it not occurred to me that he
and his chaps must be very hungry: that they must have been
growing increasingly hungry for at least this month past. They
had been engaged for six months (I don't think a single one of
them had any clear idea of time, as we at the end of countless
ages have. They still belonged to the beginnings of time—had
no inherited experience to teach them as it were), and of course,
as long as there was a piece of paper written over in accordance
with some farcical law or other made down the river, it didn't
enter anybody's head to trouble how they would live. Certainly
they had brought with them some rotten hippo-meat, which
couldn't have lasted very long, anyway, even if the pilgrims
hadn't, in the midst of a shocking hullabaloo, thrown a considerable quantity of it overboard. It looked like a high-handed
proceeding; but it was really a case of legitimate self-defence.
You can't breathe dead hippo waking, sleeping, and eating, and
at the same time keep your precarious grip on existence. Besides
that, they had given them every week three pieces of brass wire,
each about nine inches long; and the theory was they were to
buy their provisions with that currency in river-side villages.
You can see how *that* worked. There were either no villages, or
the people were hostile, or the director, who like the rest of us
fed out of tins, with an occasional old he-goat thrown in, didn't
want to stop the steamer for some more or less recondite reason.
So, unless they swallowed the wire itself, or made loops of it to

snare the fishes with, I don't see what good their extravagant salary could be to them. I must say it was paid with a regularity worthy of a large and honourable trading company. For the rest, the only thing to eat—though it didn't look eatable in the least—I saw in their possession was a few lumps of some stuff like half-cooked dough, of a dirty lavender colour, they kept wrapped in leaves, and now and then swallowed a piece of, but so small that it seemed done more for the looks of the thing than for any serious purpose of sustenance. Why in the name of all the gnawing devils of hunger they didn't go for us—they were thirty to five—and have a good tuck in for once, amazes me now when I think of it. They were big powerful men, with not much capacity to weigh the consequences, with courage, with strength, even yet, though their skins were no longer glossy and their muscles no longer hard. And I saw that something restraining, one of those human secrets that baffle probability, had come into play there. I looked at them with a swift quickening of interest—not because it occurred to me I might be eaten by them before very long, though I own to you that just then I perceived—in a new light, as it were—how unwholesome the pilgrims looked, and I hoped, yes, I positively hoped, that my aspect was not so—what shall I say?—so—unappetizing: a touch of fantastic vanity which fitted well with the dream-sensation that pervaded all my days at that time. Perhaps I had a little fever, too. One can't live with one's finger everlastingly on one's pulse. I had often 'a little fever,' or a little touch of other things—the playful paw-strokes of the wilderness, the preliminary trifling before the more serious onslaught which came in due course. Yes; I looked at them as you would on any human being, with a curiosity of their impulses, motives, capacities, weaknesses, when brought to the test of an inexorable physical necessity. Restraint! What possible restraint? Was it superstition, disgust, patience, fear—or some kind of primitive honour? No fear can stand up to hunger, no patience can wear it out, disgust simply does not exist where hunger is; and as to superstition, beliefs, and what you may call principles, they are less than chaff in a breeze. Don't you know the devilry of lingering starvation, its exasperating torment, its black thoughts, its sombre and brooding ferocity? Well, I do. It takes a man all his inborn strength to fight hunger properly. It's really easier to face bereavement, dishonour, and the perdition of one's soul—than this kind of prolonged hunger. Sad, but true. And these chaps, too, had no earthly reason for any kind of scruple. Restraint! I would just as soon have expected restraint from a hyena prowling amongst the corpses of a battlefield. But there was the fact facing me—the fact dazzling, to be seen, like the foam on the depths of the sea, like a ripple on an unfathomable enigma, a mystery greater—when I thought of it—than the curious, inexplicable note of desperate grief in this savage clamour that had swept by us on the river-bank, behind the blind whiteness of the fog.

"Two pilgrims were quarrelling in hurried whispers as to which bank. 'Left.' 'No, no; how can you? Right, right, of course.' 'It is very serious,' said the manager's voice behind me; 'I would be desolated if anything should happen to Mr. Kurtz before we came up.' I looked at him, and had not the slightest doubt he was sincere. He was just the kind of man who would wish to preserve appearances. That was his restraint. But when he muttered something about going on at once, I did not even take the trouble to answer him. I knew, and he knew, that it was impossible. Were we to let go our hold of the bottom, we would be absolutely in the air—in space. We wouldn't be able to tell where we were going to—whether up or down stream, or across—till we fetched against one bank or the other—and then we wouldn't know at first which it was. Of course I made no move. I had no mind for a smash-up. You couldn't imagine a more deadly place for a shipwreck. Whether drowned at once or not, we were sure to perish speedily in one way or another.

'I authorize you to take all the risks,' he said, after a short silence. 'I refuse to take any,' I said, shortly; which was just the answer he expected, though its tone might have surprised him. 'Well, I must defer to your judgement. You are captain,' he said, with marked civility. I turned my shoulder to him in sign of my appreciation, and looked into the fog. How long would it last? It was the most hopeless look-out. The approach to this Kurtz grubbing for ivory in the wretched bush was beset by as many dangers as though he had been an enchanted princess sleeping in a fabulous castle. 'Will they attack, do you think?' asked the manager, in a confidential tone.

"I did not think they would attack, for several obvious reasons. The thick fog was one. If they left the bank in their canoes they would get lost in it, as we would be if we attempted to move. Still, I had also judged the jungle of both banks quite impenetrable—and yet eyes were in it, eyes that had seen us. The river-side bushes were certainly very thick; but the undergrowth behind was evidently penetrable. However, during the short lift I had seen no canoes anywhere in the reach—certainly not abreast of the steamer. But what made the idea of attack inconceivable to me was the nature of the noise—of the cries we had heard. They had not the fierce character boding of immediate hostile intention. Unexpected, wild, and violent as they had been, they had given me an irresistible impression of sorrow. The glimpse of the steamboat had for some reason filled those savages with unrestrained grief. The danger, if any, I expounded, was from our proximity to a great human passion let loose. Even extreme grief may ultimately vent itself in violence—but more generally takes the form of apathy . . .

"You should have seen the pilgrims stare! They had no heart to grin, or even to revile me: but I believe they thought me gone mad—with fright, maybe. I delivered a regular lecture. My dear boys, it was no good bothering. Keep a look-out? Well, you may guess I watched the fog for the signs of lifting as a cat watches a mouse; but for anything else our eyes were of no more use to us than if we had been buried miles deep in a heap of cotton-wool. It felt like it, too—choking, warm, stifling. Besides, all I said, though it sounded extravagant, was absolutely true to fact. What we afterwards alluded to as an attack was really an attempt at repulse. The action was very far from being aggressive—it was not even defensive, in the usual sense: it was undertaken under the stress of desperation, and in its essence was purely protective.

"It developed itself, I should say, two hours after the fog lifted, and its commencement was at a spot, roughly speaking, about a mile and a half below Kurtz's station. We had just floundered and flopped round a bend, when I saw an islet, a mere grassy hummock of bright green, in the middle of the stream. It was the only thing of the kind; but as we opened the reach more, I perceived it was the head of a long sandbank, or rather of a chain of shallow patches stretching down the middle of the river. They were discoloured, just awash, and the whole lot was seen just under the water, exactly as a man's backbone is seen running down the middle of his back under the skin. Now, as far as I did see, I could go to the right or to the left of this. I didn't know either channel, of course. The banks looked pretty well alike, the depth appeared the same; but as I had been informed the station was on the west side, I naturally headed for the western passage.

"No sooner had we fairly entered it than I became aware it was much narrower than I had supposed. To the left of us there was the long uninterrupted shoal, and to the right a high, steep bank heavily overgrown with bushes. Above the bush the trees stood in serried ranks. The twigs overhung the current thickly, and from distance to distance a large limb of some tree projected rigidly over the stream. It was then well on in the afternoon, the face of the forest was gloomy, and a broad strip of shadow had already fallen on the water. In this shadow we steamed up—very slowly, as you may imagine. I sheered her well inshore—the water being deepest near the bank, as the sounding-pole informed me.

"One of my hungry and forbearing friends was sounding in the bows just below me. This steamboat was exactly like a decked scow. On the deck, there were two little teak-wood houses, with doors and windows. The boiler was in the fore-end, and the machinery right astern. Over the whole there was a light roof, supported on stanchions. The funnel projected through that roof, and in front of the funnel a small cabin built of light planks served for a pilot-house. It contained a couch, two camp-stools, a loaded Martini-Henry leaning in one corner, a tiny table, and the steering-wheel. It had a wide door in front and a broad shutter at each side. All these were always thrown open, of course. I spent my days perched up there on the extreme fore-end of that roof, before the door. At night I slept, or tried to, on the couch. An athletic black belonging to some coast tribe, and educated by my poor predecessor, was the helmsman. He sported a pair of brass earrings, wore a blue cloth wrapper from the waist to the ankles, and thought all the world of himself. He was the most unstable kind of fool I had ever seen. He steered with no end of a swagger while you were by; but if he lost sight of you, he became instantly the prey of an abject funk, and would let that cripple of a steamboat get the upper hand of him in a minute.

"I was looking down at the sounding-pole, and feeling much annoyed to see at each try a little more of it stick out of that river, when I saw my poleman give up the business suddenly, and stretch himself flat on the deck, without even taking the trouble to haul his pole in. He kept hold on it though, and it trailed in the water. At the same time the fireman, whom I could also see below me, sat down abruptly before his furnace and ducked his head. I was amazed. Then I had to look at the river mighty quick, because there was a snag in the fairway.

Sticks, little sticks, were flying about—thick: they were whizzing before my nose, dropping below me, striking behind me against my pilot-house. All this time the river, the shore, the woods, were very quiet—perfectly quiet. I could only hear the heavy splashing thump of the stern-wheel and the patter of these things. We cleared the snag clumsily. Arrows, by Jove! We were being shot at! I stepped in quickly to close the shutter on the land-side. That fool-helmsman, his hands on the spokes, was lifting his knees high, stamping his feet, champing his mouth, like a reined-in horse. Confound him! And we were staggering within ten feet of the bank. I had to lean right out to swing the heavy shutter, and I saw a face amongst the leaves on the level with my own, looking at me very fierce and steady and then suddenly, as though a veil had been removed from my eyes, I made out, deep in the tangled gloom, naked breasts, arms, legs, glaring eyes—the bush was swarming with human limbs in movement, glistening, of bronze colour. The twigs shook, swayed, and rustled, the arrows flew out of them, and then the shutter came to. 'Steer her straight,' I said to the helmsman. He held his head rigid, face forward; but his eyes rolled, he kept on, lifting and setting down his feet gently, his mouth foamed a little. 'Keep quiet!' I said in a fury. I might just as well have ordered a tree not to sway in the wind. I darted out. Below me there was a great scuffle of feet on the iron deck; confused exclamations; a voice screamed, 'Can you turn back?' I caught sight of a V-shaped ripple on the water ahead. What? Another snag! A fusillade burst out under my feet. The pilgrims had opened with their Winchesters, and were simply squirting lead into that bush. A deuce of a lot of smoke came up and drove slowly forward. I swore at it. Now I couldn't see the ripple or the snag either. I stood in the doorway, peering, and the arrows came in swarms. They might have been poisoned, but they looked as though they wouldn't kill a cat. The bush began to howl. Our wood-cutters raised a warlike whoop; the report of a rifle just at my back deafened me. I glanced over my shoulder, and the pilot-house was yet full of noise and smoke when I made a dash at the wheel. The fool-nigger had dropped everything, to throw the shutter open and let off that Martini-Henry. He stood before the wide opening, glaring, and I yelled at him to come back, while I straightened the sudden twist out of that steamboat. There was no room to turn even if I had wanted to, the snag was somewhere very near ahead in that confounded smoke, there was no time to lose, so I just crowded her into the bank—right into the bank, where I knew the water was deep.

"We tore slowly along the overhanging bushes in a whirl of broken twigs and flying leaves. The fusillade below stopped short, as I had foreseen it would when the squirts got empty. I threw my head back to a glinting whizz that traversed the pilot-house, in at one shutterhole and out at the other. Looking past that mad helmsman, who was shaking the empty rifle and yelling at the shore, I saw vague forms of men running bent double, leaping, gliding, distinct, incomplete, evanescent. Something big appeared in the air before the shutter, the rifle went overboard, and the man stepped back swiftly, looked at me over his shoulder in an extraordinary, profound, familiar manner, and fell upon my feet. The side of his head hit the wheel twice, and the end of what appeared a long cane clattered round and knocked over a little camp-stool. It looked as though after wrenching that thing from somebody ashore he had lost his balance in the effort. The thin smoke had blown away, we were clear of the snag, and looking ahead I could see that in another hundred yards or so I would be free to sheer off, away from the bank; but my feet felt so very warm and wet that I had to look down. The man had rolled on his back and stared straight up at me; both his hands clutched that cane. It was the shaft of a spear that, either thrown or lunged through the opening, had caught him in the side just below the ribs; the blade had gone in out of sight, after making a frightful gash; my shoes were full; a pool of blood lay very still, gleaming dark-red under the wheel; his eyes shone with an amazing lustre. The fusillade burst out again. He looked at me anxiously, gripping the spear like something precious, with an air of being afraid I would try to take it away from him. I had to make an effort to free my eyes from his gaze and attend to the steering. With one hand I felt above my head for the line of the steam whistle, and jerked out screech after screech hurriedly. The tumult of angry and warlike yells was checked instantly, and then from the depths of the woods went out such a tremulous and prolonged wail of mournful fear and utter despair as may be imagined to follow the flight of the last hope from the earth. There was a great commotion in the bush; the shower of arrows stopped, a few dropping shots rang out sharply—then silence, in which the languid beat of the stern-wheel came plainly to my ears. I put the helm hard a-starboard at the moment when the pilgrim in pink pyjamas, very hot and agitated, appeared in the doorway. 'The manager sends me—' he began in an official tone, and stopped short. 'Good God!' he said, glaring at the wounded man.

"We two whites stood over him, and his lustrous and inquiring glance enveloped us both. I declare it looked as though he would presently put to us some question in an understandable language; but he died without uttering a sound, without moving a limb, without twitching a muscle. Only in the very last moment, as though in response to some sign we could not see, to some whisper we could not hear, he frowned heavily, and that frown gave to his black death-mask an inconceivably sombre, brooding, and menacing expression. The lustre of inquiring glance faded swiftly into vacant glassiness. 'Can you steer?' I asked the agent eagerly. He looked very dubious; but I made a grab at his arm, and he understood at once I meant him to steer whether or no. To tell you the truth, I was morbidly anxious to change my shoes and socks. 'He is dead,' murmured the fellow, immensely impressed. 'No doubt about it,' said I, tugging like mad at the shoe-laces. 'And by the way, I suppose Mr. Kurtz is dead as well by this time.'

"For the moment that was the dominant thought. There was a sense of extreme disappointment, as though I had found out I had been striving after something altogether without a substance. I couldn't have been more disgusted if I had travelled all this way for the sole purpose of talking with Mr. Kurtz. Talking with . . . I flung one shoe overboard, and became aware that that was exactly what I had been looking forward to—a talk with Kurtz. I made the strange discovery that I had never imagined him as doing, you know, but as discoursing. I didn't say to myself, 'Now I will never see him,' or 'Now I will never shake him by the hand,' but, 'Now I will never hear him.' The man presented himself as a voice. Not of course that I did not connect him with some sort of action. Hadn't I been told in all the tones of jealousy and admiration that he had collected, bartered, swindled, or stolen more ivory than all the other agents together? That was not the point. The point was in his being a gifted creature, and that of all his gifts the one that stood out pre-eminently, that carried with it a sense of real presence, was his ability to talk, his words—the gift of expression, the bewildering, the illuminating, the most exalted and the most contemptible, the pulsating stream of light, or the deceitful flow from the heart of an impenetrable darkness.

"The other shoe went flying unto the devil-god of that river. I thought, By Jove! it's all over. We are too late; he has vanished—the gift has vanished, by means of some spear, arrow, or club. I will never hear that chap speak after all—and my sorrow had a startling extravagance of emotion, even such as I had noticed in the howling sorrow of these savages in the bush. I couldn't have felt more of lonely desolation somehow, had I been robbed of a belief or had missed my destiny in life . . . Why do you sigh in this beastly way, somebody? Absurd? Well, absurd. Good Lord! mustn't a man ever—Here, give me some tobacco.''. . .

There was a pause of profound stillness, then a match flared, and Marlow's lean face appeared, worn, hollow, with downward folds and dropped eyelids, with an aspect of concentrated attention; and as he took vigorous draws at his pipe, it seemed to retreat and advance out of the night in the regular flicker of the tiny flame. The match went out.

"Absurd!" he cried. "This is the worst of trying to tell. . . . Here you all are, each moored with two good addresses, like a hulk with two anchors, a butcher round one corner, a policeman round another, excellent appetites, and temperature normal—you hear—normal from year's end to year's end. And you say, Absurd! Absurd be—exploded! Absurd! My dear boys, what can you expect from a man who out of sheer nervousness had just flung overboard a pair of new shoes! Now I think of it, it is amazing I did not shed tears. I am, upon the whole, proud of my fortitude. I was cut to the quick at the idea of having lost the inestimable privilege of listening to the gifted Kurtz. Of course I was wrong. The privilege was waiting for me. Oh, yes, I heard more than enough. And I was right too. A voice. He was very little more than a voice. And I heard—him—it—this voice—other voices—all of them were so little more than voices—and the memory of that time itself lingers around me, impalpable, like a dying vibration of one immense jabber, silly, atrocious, sordid, savage, or simply mean, without any kind of sense. Voices, voices—even the girl herself—now—"

He was silent for a long time.

"I laid the ghost of his gifts at last with a lie," he began, suddenly. "Girl! What? Did I mention a girl? Oh, she is out of it—completely. They—the women I mean—are out of it—should be out of it. We must help them to stay in that beautiful world of their own, lest ours gets worse. Oh, she had to be out of it. You should have heard the disinterred body of Mr. Kurtz saying, 'My Intended.' You would have perceived directly then how completely she was out of it. And the lofty frontal bone of Mr. Kurtz! They say the hair goes on growing sometimes, but this—ah—specimen, was impressively bald. The wilderness had patted him on the head, and, behold, it was like a ball— an ivory ball; it had caressed him, and—lo!—he had withered; it had taken him, loved him, embraced him, got into his veins, consumed his flesh, and sealed his soul to its own by the inconceivable ceremonies of some devilish initiation. He was its spoiled and pampered favourite. Ivory? I should think so. Heaps of it, stacks of it. The old mud shanty was bursting with it. You would think there was not a single tusk left either above or below the ground in the whole country. 'Mostly fossil,' the manager had remarked, disparagingly. It was no more fossil than I am; but they call it fossil when it is dug up. It appears these niggers do bury the tusks sometimes—but evidently they couldn't bury this parcel deep enough to save the gifted Mr. Kurtz from his fate. We filled the steamboat with it, and had to pile a lot on the deck. Thus he could see and enjoy as long as he could see, because the appreciation of this favour had remained with him to the last. You should have heard him say, 'My ivory.' Oh yes, I heard him. 'My Intended, my ivory, my station, my river, my—' everything belonged to him. It made me hold my breath in expectation of hearing the wilderness burst into a prodigious peal of laughter that would shake the fixed stars in their places. Everything belonged to him—but that was a trifle. The thing was to know what he belonged to, how many powers of darkness claimed him for their own. That was the reflection that made you creepy all over. It was impossible —it was not good for one either—trying to imagine. He had taken a high seat amongst the devils of the land—I mean literally. You can't understand. How could you?—with solid pavement under your feet, surrounded by kind neighbours ready to cheer you or to fall on you, stepping delicately between the

butcher and the policeman, in the holy terror of scandal and
gallows and lunatic asylums—how can you imagine what particular region of the first ages a man's untrammelled feet may
take him into by the way of solitude—utter solitude without a
policeman—by the way of silence—utter silence, where no warning voice of a kind neighbour can be heard whispering of public
opinion? These little things make all the great difference. When
they are gone you must fall back upon your own innate strength,
upon your own capacity for faithfulness. Of course you may be
too much of a fool to go wrong—too dull even to know you
are being assaulted by the powers of darkness. I take it, no fool
ever made a bargain for his soul with the devil: the fool is too
much of a fool, or the devil too much of a devil—I don't know
which. Or you may be such a thunderingly exalted creature as
to be altogether deaf and blind to anything but heavenly sights
and sounds. Then the earth for you is only a standing place—
and whether to be like this is your loss or your gain I won't
pretend to say. But most of us are neither one nor the other.
The earth for us is a place to live in, where we must put up
with sights, with sounds, with smells, too, by Jove!—breathe
dead hippo, so to speak, and not be contaminated. And there,
don't you see? your strength comes in, the faith in your ability
for the digging of unostentatious holes to bury the stuff in—
your power of devotion, not to yourself, but to an obscure, back-
breaking business. And that's difficult enough. Mind, I am not
trying to excuse or even explain—I am trying to account to
myself for—for—Mr. Kurtz—for the shade of Mr. Kurtz. This
initiated wraith from the back of Nowhere honoured me with
its amazing confidence before it vanished altogether. This was
because it could speak English to me. The original Kurtz had
been educated partly in England, and—as he was good enough
to say himself—his sympathies were in the right place. His
mother was half-English, his father was half-French. All Europe
contributed to the making of Kurtz; and by-and-by I learned

that, most appropriately, the International Society for the Suppression of Savage Customs had intrusted him with the making of a report, for its future guidance. And he had written it, too. I've seen it. I've read it. It was eloquent, vibrating with eloquence, but too high-strung, I think. Seventeen pages of close writing he had found time for! But this must have been before his—let us say—nerves, went wrong, and caused him to preside at certain midnight dances ending with unspeakable rites, which—as far as I reluctantly gathered from what I heard at various times—were offered up to him—do you understand?— to Mr. Kurtz himself. But it was a beautiful piece of writing. The opening paragraph, however, in the light of later infor- mation, strikes me now as ominous. He began with the argu- ment that we whites, from the point of development we had arrived at, 'must necessarily appear to them [savages] in the nature of supernatural beings—we approach them with the might as of a deity,' and so on, and so on. 'By the simple exercise of our will we can exert a power for good practically unbounded,' etc., etc. From that point he soared and took me with him. The peroration was magnificent, though difficult to remember, you know. It gave me the notion of an exotic Immensity ruled by an august Benevolence. It made me tingle with enthusiasm. This was the unbounded power of eloquence—of words—of burning noble words. There were no practical hints to interrupt the magic current of phrases, unless a kind of note at the foot of the last page, scrawled evidently much later, in an unsteady hand, may be regarded as the exposition of a method. It was very simple, and at the end of that moving appeal to every altruistic sentiment it blazed at you, luminous and terrifying, like a flash of lightning in a serene sky: 'Exterminate all the brutes!' The curious part was that he had apparently forgotten all about that valuable postscriptum, because, later on, when he in a sense came to himself, he repeatedly entreated me to take good care of 'my pamphlet' (he called it), as it was sure to have in the future a good influence upon his career. I had full information about all these things, and, besides, as it turned out, I was to have the care of his memory. I've done enough for it to give the indisputable right to lay it, if I choose, for an everlasting rest in the dust-bin of progress, amongst all the sweepings and, figuratively speaking, all the dead cats of civilization. But then, you see, I can't choose. He won't be forgotten. Whatever he was, he was not common. He had the power to charm or frighten rudimentary souls into an aggravated witch-dance in his honour; he could also fill the small souls of the pilgrims with bitter misgivings: he had one devoted friend at least, and he had conquered one soul in the world that was neither rudimentary nor tainted with self-seeking. No; I can't forget him, though I am not prepared to affirm the fellow was exactly worth the life we lost in getting to him. I missed my late helmsman awfully—I missed him even while his body was still lying in the pilot-house. Perhaps you will think it passing strange this regret for a savage who was no more account than a grain of sand in a black Sahara. Well, don't you see, he had done something, he had steered; for months I had him at my back—a help—an instrument. It was a kind of partnership. He steered for me—I had to look after him, I worried about his deficiencies, and thus a subtle bond had been created, of which I only became aware when it was suddenly broken. And the intimate profundity of that look he gave me when he received his hurt remains to this day in my memory—like a claim of distant kinship affirmed in a supreme moment.

"Poor fool! If he had only left that shutter alone. He had no restraint, no restraint—just like Kurtz—a tree swayed by the wind. As soon as I had put on a dry pair of slippers, I dragged him out, after first jerking the spear out of his side, which operation I confess I performed with my eyes shut tight. His heels leaped together over the little doorstep; his shoulders were pressed to my breast; I hugged him from behind desperately.

Oh! he was heavy, heavy; heavier than any man on earth, I should imagine. Then without more ado I tipped him overboard. The current snatched him as though he had been a wisp of grass, and I saw the body roll over twice before I lost sight of it for ever. All the pilgrims and the manager were then congregated on the awning-deck about the pilot-house, chattering at each other like a flock of excited magpies, and there was a scandalized murmur at my heartless promptitude. What they wanted to keep that body hanging about for I can't guess. Embalm it, maybe. But I had also heard another, and a very ominous, murmur on the deck below. My friends the woodcutters were likewise scandalized, and with a better show of reason—though I admit that the reason itself was quite inadmissible. Oh, quite! I had made up my mind that if my late helmsman was to be eaten, the fishes alone should have him. He had been a very second-rate helmsman while alive, but now he was dead he might have become a first-class temptation, and possibly cause some startling trouble. Besides, I was anxious to take the wheel, the man in pink pyjamas showing himself a hopeless duffer at the business.

"This I did directly the simple funeral was over. We were going half-speed, keeping right in the middle of the stream, and I listened to the talk about me. They had given up Kurtz, they had given up the station; Kurtz was dead, and the station had been burnt—and so on—and so on. The red-haired pilgrim was beside himself with the thought that at least this poor Kurtz had been properly avenged. 'Say! We must have made a glorious slaughter of them in the bush. Eh? What do you think? Say?' He positively danced, the bloodthirsty little gingery beggar. And he had nearly fainted when he saw the wounded man! I could not help saying, 'You made a glorious lot of smoke, anyhow.' I had seen, from the way the tops of the bushes rustled and flew, that almost all the shots had gone too high. You can't hit anything unless you take aim and fire from the shoulder; but these chaps fired from the hip with their eyes shut. The retreat, I maintained—and I was right—was caused by the screeching of the steam-whistle. Upon this they forgot Kurtz, and began to howl at me with indignant protests.

"The manager stood by the wheel murmuring confidentially about the necessity of getting well away down the river before dark at all events, when I saw in the distance a clearing on the river-side and the outlines of some sort of building. 'What's this?' I asked. He clapped his hands in wonder. 'The station!' he cried. I edged in at once, still going half-speed.

"Through my glasses I saw the slope of a hill interspersed with rare trees and perfectly free from undergrowth. A long decaying building on the summit was half buried in the high grass; the large holes in the peaked roof gaped black from afar; the jungle and the woods made a background. There was no enclosure or fence of any kind; but there had been one apparently, for near the house half-a-dozen slim posts remained in a row, roughly trimmed, and with their upper ends ornamented with round carved balls. The rails, or whatever there had been between, had disappeared. Of course the forest surrounded all that. The river-bank was clear, and on the water-side I saw a white man under a hat like a cart-wheel beckoning persistently with his whole arm. Examining the edge of the forest above and below, I was almost certain I could see movements—human forms gliding here and there. I steamed past prudently, then stopped the engines and let her drift down. The man on the shore began to shout, urging us to land. 'We have been attacked,' screamed the manager. 'I know—I know. It's all right,' yelled back the other, as cheerful as you please. 'Come along. It's all right. I am glad.'

"His aspect reminded me of something I had seen—something funny I had seen somewhere. As I manoeuvred to get alongside, I was asking myself, 'What does this fellow look like?' Suddenly I got it. He looked like a harlequin. His clothes had been made of some stuff that was brown holland probably, but it was covered with patches all over, with bright patches, blue, red, and yellow—patches on the back, patches on the front, patches on elbows, on knees; coloured binding around his jacket, scarlet edging at the bottom of his trousers; and the sunshine made him look extremely gay and wonderfully neat withal, because you would see how beautifully all this patching had been done. A beardless, boyish face, very fair, no features to speak of, nose peeling, little blue eyes, smiles and frowns chasing each other over that open countenance like sunshine and shadow on a wind-swept plain. 'Look out, captain!' he cried; 'there's a snag lodged in here last night.' What! Another snag? I confess I swore shamefully. I had nearly holed my cripple, to finish off that charming trip. The harlequin on the bank turned his little pug-nose up to me. 'You English?' he asked, all smiles. 'Are you?' I shouted from the wheel. The smiles vanished, and he shook his head as if sorry for my disappointment. Then he brightened up. 'Never mind!' he cried, encouragingly. 'Are we in time?' I asked. 'He is up there,' he replied, with a toss of the head up the hill, and becoming gloomy all of a sudden. His face was like the autumn sky, overcast one moment and bright the next.

"When the manager, escorted by the pilgrims, all of them armed to the teeth, had gone to the house this chap came on board. 'I say, I don't like this. These natives are in the bush,' I said. He assured me earnestly it was all right. 'They are simple people,' he added; 'well, I am glad you came. It took me all my time to keep them off.' 'But you said it was all right,' I cried. 'Oh, they meant no harm,' he said; and as I stared he corrected himself. 'Not exactly.' Then vivaciously, 'My faith, your pilot-house wants a clean up!' In the next breath he advised me to keep enough steam on the boiler to blow the whistle in case of any trouble. 'One good screech will do more for you than all your rifles. They are simple people,' he repeated. He rattled away at such a rate he quite overwhelmed me. He seemed to be trying to make up for lots of silence, and actually hinted, laughing, that such was the case. 'Don't you talk with Mr. Kurtz?' I said. 'You don't talk with that man—you listen to him,' he exclaimed with severe exaltation. 'But now—' He waved his arm, and in the twinkling of an eye was in the uttermost depths of despondency. In a moment he came up again with a jump, possessed himself of both my hands, shook them continuously, while he gabbled: 'Brother sailor . . . honour . . . pleasure . . . delight . . . introduce myself . . . Russian . . . son of an arch-priest . . . Government of Tambov . . . What? Tobacco! English tobacco; the excellent English tobacco! Now, that's brotherly. Smoke? Where's a sailor that does not smoke?'

"The pipe soothed him, and gradually I made out he had run away from school, had gone to sea in a Russian ship; ran away again; served some time in English ships; was now reconciled with the arch-priest. He made a point of that. 'But when one is young one must see things, gather experience, ideas; enlarge the mind.' 'Here!' I interrupted. 'You can never tell! Here I met Mr. Kurtz,' he said, youthfully solemn and reproachful. I held my tongue after that. It appears he had persuaded a Dutch trading-house on the coast to fit him out with stores and goods, and had started for the interior with a light heart, and no more idea of what would happen to him than a baby. He had been wandering about that river for nearly two years alone, cut off from everybody and everything. 'I am not so young as I look. I am twenty-five,' he said. 'At first old Van Shuyten would tell me to go to the devil,' he narrated with keen enjoyment; 'but I stuck to him, and talked and talked, till at last he got afraid I would talk the hind-leg off his favourite dog, so he gave me some cheap things and a few guns, and told me he hoped he would never see my face again. Good old Dutchman, Van Shuyten. I've sent him one small lot of ivory a year ago, so that he can't call me a little thief when I get

back. I hope he got it. And for the rest I don't care. I had some
wood stacked for you. That was my old house. Did you see?'

"I gave him Towson's book. He made as though he would
kiss me, but restrained himself. 'The only book I had left, and
I thought I had lost it,' he said, looking at it ecstatically. 'So
many accidents happen to a man going about alone, you know.
Canoes get upset sometimes—and sometimes you've got to clear
out so quick when the people get angry.' He thumbed the pages.
'You made notes in Russian?' I asked. He nodded. 'I thought
they were written in cipher,' I said. He laughed, then became
serious. 'I had lots of trouble to keep these people off,' he said.
'Did they want to kill you?' I asked. 'Oh, no!' he cried, and
checked himself. 'Why did they attack us?' I pursued. He hes-
itated, then said shamefacedly, 'They don't want him to go.'
'Don't they?' I said, curiously. He nodded a nod full of mystery
and wisdom. 'I tell you,' he cried, 'this man has enlarged my
mind.' He opened his arms wide, staring at me with his little
blue eyes that were perfectly round."

3

"I looked at him, lost in astonishment. There he was before
me, in motley, as though he had absconded from a troupe of
mimes, enthusiastic, fabulous. His very existence was improb-
able, inexplicable, and altogether bewildering. He was an insoluble problem. It was inconceivable how he had existed, how
he had succeeded in getting so far, how he had managed to
remain—why he did not instantly disappear. 'I went a little
farther,' he said, 'then still a little farther—till I had gone so
far that I don't know how I'll ever get back. Never mind. Plenty
time. I can manage. You take Kurtz away quick—quick—I tell
you.' The glamour of youth enveloped his parti-coloured rags,
his destitution, his loneliness, the essential desolation of his futile
wanderings. For months—for years—his life hadn't been worth
a day's purchase; and there he was gallantly, thoughtlessly alive,

to all appearance indestructible solely by the virtue of his few years and of his unreflecting audacity. I was seduced into something like admiration—like envy. Glamour urged him on, glamour kept him unscathed. He surely wanted nothing from the wilderness but space to breathe in and to push on through. His need was to exist, and to move onwards at the greatest possible risk, and with a maximum of privation. If the absolutely pure, uncalculating, unpractical spirit of adventure had ever ruled a human being, it ruled this be-patched youth. I almost envied him the possession of this modest and clear flame. It seemed to have consumed all thought of self so completely, that even while he was talking to you, you forget that it was he—the man before your eyes—who had gone through these things. I did not envy him his devotion to Kurtz, though. He had not meditated over it. It came to him, and he accepted it with a sort of eager fatalism. I must say that to me it appeared about the most dangerous thing in every way he had come upon so far.

"They had come together unavoidably, like two ships becalmed near each other, and lay rubbing sides at last. I suppose Kurtz wanted an audience, because on a certain occasion, when encamped in the forest, they had talked all night, or more probably Kurtz had talked. 'We talked of everything,' he said, quite transported at the recollection. 'I forgot there was such a thing as sleep. The night did not seem to last an hour. Everything! Everything! . . . Of love, too.' 'Ah, he talked to you of love!' I said, much amused. 'It isn't what you think,' he cried, almost passionately. 'It was in general. He made me see things—things.'

"He threw his arms up. We were on deck at the time, and the headman of my wood-cutters, lounging near by, turned upon him his heavy and glittering eyes. I looked around, and I don't know why, but I assure you that never, never before, did this land, this river, this jungle, the very arch of this blazing sky, appear to me so hopeless and so dark, so impenetrable to human thought, so pitiless to human weakness. 'And, ever since, you have been with him, of course?' I said.

"On the contrary. It appears their intercourse had been very much broken by various causes. He had, as he informed me proudly, managed to nurse Kurtz through two illnesses (he alluded to it as you would to some risky feat), but as a rule Kurtz wandered alone, far in the depths of the forest. 'Very often coming to this station, I had to wait days and days before he would turn up,' he said. 'Ah, it was worth waiting for!—sometimes.' 'What was he doing? exploring or what?' I asked. 'Oh, yes, of course'; he had discovered lots of villages, a lake, too— he did not know exactly in what direction; it was dangerous to inquire too much—but mostly his expeditions had been for ivory. 'But he had no goods to trade with by that time,' I objected. 'There's a good lot of cartridges left even yet,' he answered, looking away. 'To speak plainly, he raided the country,' I said. He nodded. 'Not alone, surely!' He muttered something about the villages round that lake. 'Kurtz got the tribe to follow him, did he?' I suggested. He fidgeted a little. 'They adored him,' he said. The tone of these words was so extraordinary that I looked at him searchingly. It was curious to see his mingled eagerness and reluctance to speak of Kurtz. The man filled his life, occupied his thoughts, swayed his emotions. 'What can you expect?' he burst out; 'he came to them with thunder and lightning, you know—and they had never seen anything like it—and very terrible. He could be very terrible. You can't judge Mr. Kurtz as you would an ordinary man. No, no, no! Now—just to give you an idea—I don't mind telling you, he wanted to shoot me, too, one day—but I don't judge him.' 'Shoot you!' I cried. 'What for?' 'Well, I had a small lot of ivory the chief of that village near my house gave me. You see I used to shoot game for them. Well, he wanted it, and wouldn't hear reason. He declared he would shoot me unless I gave him the ivory and then cleared out of the country, because he could do so, and had a fancy for it, and there was nothing on earth to prevent him killing whom he jolly well pleased. And it was true, too. I gave him the ivory. What did I care! But I didn't clear out. No, no. I couldn't leave him. I had to be careful, of course, till we got friendly again for a time. He had his second illness then. Afterwards I had to keep out of the way; but I didn't mind. He was living for the most part in those villages on the lake. When he came down to the river, sometimes he would take to me, and sometimes it was better for me to be careful. This man suffered too much. He hated all this, and somehow he couldn't get away. When I had a chance I begged him to try and leave while there was time; I offered to go back with him. And he would say yes, and then he would remain; go off on another ivory hunt; disappear for weeks; forget himself amongst these people—forget himself—you know.' 'Why! he's mad,' I said. He protested indignantly. Mr. Kurtz couldn't be mad. If I had heard him talk, only two days ago, I wouldn't dare hint at such a thing ... I had taken up my binoculars while we talked, and was looking at the shore, sweeping the limit of the forest at each side and at the back of the house. The consciousness of there being people in that bush, so silent, so quiet—as silent and quiet as the ruined house on the hill— made me uneasy. There was no sign on the face of nature of this amazing tale that was not so much told as suggested to me in desolate exclamations, completed by shrugs, in interrupted phrases, in hints ending in deep sighs. The woods were un- moved, like a mask—heavy, like the closed door of a prison— they looked with their air of hidden knowledge, of patient expectation, of unapproachable silence. The Russian was ex- plaining to me that it was only lately that Mr. Kurtz had come down the river, bringing along with him all the fighting men of that lake tribe. He had been absent for several months— getting himself adored, I suppose—and had come down un- expectedly, with the intention to all appearance of making a raid either across the river or down stream. Evidently the appetite for more ivory had got the better of the—what shall I say?—less material aspirations. However he had got much worse suddenly. 'I heard he was lying helpless, and so I came up—took my chance,' said the Russian. 'Oh, he is bad, very bad.' I directed my glass to the house. There were no signs of life, but there was the ruined roof, the long mud wall peeping above the grass, with three little square window-holes, no two the same size: all this brought within reach of my hand, as it were. And then I made a brusque movement, and one of the remaining posts of that vanished fence leaped up in the field of my glass. You remember I told you I had been struck at the distance by certain attempts at ornamentation, rather remarkable in the ruinous aspect of the place. Now I had suddenly a nearer view, and its first result was to make me throw my head back as if before a blow. Then I went carefully from post to post with my glass, and I saw my mistake. These round knobs were not ornamental but symbolic; they were expressive and puzzling, striking and disturbing—food for thought and also for the vultures if there had been any looking down from the sky; but at all events for such ants as were industrious enough to ascend the pole. They would have been even more impressive, those heads on the stakes, if their faces had not been turned to the house. Only one, the first I had made out, was facing my way. I was not so shocked as you may think. The start back I had given was really nothing but a movement of surprise. I had expected to see a knob of wood there, you know. I returned deliberately to the first I had seen—and there it was, black, dried, sunken, with closed eyelids—a head that seemed to sleep at the top of that pole, and, with the shrunken dry lips showing a narrow white line of the teeth, was smiling, too, smiling continuously at some endless and jocose dream of that eternal slumber.

"I am not disclosing any trade secrets. In fact, the manager said afterwards that Mr. Kurtz's methods had ruined the district.

I have no opinion on that point, but I want you clearly to understand that there was nothing exactly profitable in these heads being there. They only showed that Mr. Kurtz lacked restraint in the gratification of his various lusts, that there was something wanting in him—some small matter which, when the pressing need arose, could not be found under his magnificent eloquence. Whether he knew of this deficiency himself I can't say. I think the knowledge came to him at last—only at the very last. But the wilderness had found him out early, and had taken on him a terrible vengeance for the fantastic invasion. I think it had whispered to him things about himself which he did not know, things of which he had no conception till he took counsel with this great solitude—and the whisper had proved irresistibly fascinating. It echoed loudly within him because he was hollow at the core . . . I put down the glass, and the head that had appeared near enough to be spoken to seemed at once to have leaped away from me into inaccessible distance.

"The admirer of Mr. Kurtz was a bit crestfallen. In a hurried, indistinct voice he began to assure me he had not dared to take these—say, symbols—down. He was not afraid of the natives; they would not stir till Mr. Kurtz gave the word. His ascendancy was extraordinary. The camps of these people surrounded the place, and the chiefs came every day to see him. They would crawl . . . 'I don't want to know anything of the ceremonies used when approaching Mr. Kurtz,' I shouted. Curious, this feeling that came over me that such details would be more intolerable than those heads drying on the stakes under Mr. Kurtz's windows. After all, that was only a savage sight, while I seemed at one bound to have been transported into some lightless region of subtle horrors, where pure, uncomplicated savagery was a positive relief, being something that had a right to exist—obviously—in the sunshine. The young man looked at me with surprise. I suppose it did not occur to him that Mr. Kurtz was no idol of mine. He forgot I hadn't heard any of these splendid monologues on, what was it? on love, justice, conduct of life—or what not. If it had come to crawling before Mr. Kurtz, he crawled as much as the veriest savage of them all. I had no idea of the conditions, he said: these heads were the heads of rebels. I shocked him excessively by laughing. Rebels! What would be the next definition I was to hear? There had been enemies, criminals, workers—and these were rebels. Those rebellious heads looked very subdued to me on their sticks. 'You don't know how such a life tries a man like Kurtz,' cried Kurtz's last disciple. 'Well, and you?' I said. 'I! I! I am a simple man. I have no great thoughts. I want nothing from anybody. How can you compare me to? . . .' His feelings were too much for speech, and suddenly he broke down. 'I don't understand,' he groaned. 'I've been doing my best to keep him alive, and that's enough. I had no hand in all this. I have no abilities. There hasn't been a drop of medicine or a mouthful of invalid food for months here. He was shamefully abandoned. A man like this, with such ideas. Shamefully! Shamefully! I—I—haven't slept for the last ten nights . . .'

"His voice lost itself in the calm of the evening. The long shadows of the forest had slipped down hill while we talked, had gone far beyond the ruined hovel, beyond the symbolic row of stakes. All this was in the gloom, while we down there were yet in the sunshine, and the stretch of the river abreast of the clearing glittered in a still and dazzling splendour, with a murky and overshadowed bend above and below. Not a living soul was seen on the shore. The bushes did not rustle.

"Suddenly round the corner of the house a group of men appeared, as though they had come up from the ground. They waded waist-deep in the grass, in a compact body, bearing an improvised stretcher in their midst. Instantly, in the emptiness of the landscape, a cry arose whose shrillness pierced the still air like a sharp arrow flying straight to the very heart of the land; and, as if by enchantment, streams of human beings—of naked human beings—with spears in their hands, with bows, with shields, with wild glances and savage movements, were poured into the clearing by the darkfaced and pensive forest. The bushes shook, the grass swayed for a time, and then everything stood still in attentive immobility.

" 'Now, if he does not say the right thing to them we are all done for,' said the Russian at my elbow. The knot of men with the stretcher had stopped, too, halfway to the steamer, as if petrified. I saw the man on the stretcher sit up, lank and with an uplifted arm, above the shoulders of the bearers. 'Let us hope that the man who can talk so well of love in general will find some particular reason to spare us this time,' I said. I resented bitterly the absurd danger of our situation, as if to be at the mercy of that atrocious phantom had been a dishonouring necessity. I could not hear a sound, but through my glasses I saw the thin arm extended commandingly, the lower jaw moving, the eyes of that apparition shining darkly far in its bony head that nodded with grotesque jerks. Kurtz—Kurtz—that means short in German—don't it? Well, the name was as true as everything else in his life—and death. He looked at least seven feet long. His covering had fallen off, and his body emerged from it pitiful and appalling as from a winding-sheet. I could see the cage of his ribs all astir, the bones of his arm waving. It was as though an animated image of death carved out of old ivory had been shaking its hand with menaces at a motionless crowd of men made of dark and glittering bronze. I saw him open his mouth wide—it gave him a weirdly voracious aspect, as though he had wanted to swallow all the air, all the earth, all the men before him. A deep voice reached me faintly. He must have been shouting. He fell back suddenly. The stretcher shook as the bearers staggered forward again, and almost at the same time I noticed that the crowd of savages was vanishing without any perceptible movement of retreat, as if the forest that had ejected these beings so suddenly had drawn them in again as the breath is drawn in a long aspiration.

"Some of the pilgrims behind the stretcher carried his arms—two shot-guns, a heavy rifle, and a light revolver-carbine—the thunderbolts of that pitiful Jupiter. The manager bent over him murmuring as he walked beside his head. They laid him down in one of the little cabins—just a room for a bedplace and a camp-stool or two, you know. We had brought his belated correspondence, and a lot of torn envelopes and open letters littered his bed. His hand roamed feebly amongst these papers. I was struck by the fire of his eyes and the composed languor of his expression. It was not so much the exhaustion of disease. He did not seem in pain. This shadow looked satiated and calm, as though for the moment it had had its fill of all the emotions.

"He rustled one of the letters, and looking straight in my face said, 'I am glad.' Somebody had been writing to him about me. These special recommendations were turning up again. The volume of tone he emitted without effort, almost without the trouble of moving his lips, amazed me. A voice! a voice! It was grave, profound, vibrating, while the man did not seem capable of a whisper. However, he had enough strength in him—factitious no doubt—to very nearly make an end of us, as you shall hear directly.

"The manager appeared silently in the doorway; I stepped out at once and he drew the curtain after me. The Russian, eyed curiously by the pilgrims, was staring at the shore. I followed the direction of his glance.

"Dark human shapes could be made out in the distance, flitting indistinctly against the gloomy border of the forest, and near the river two bronze figures, leaning on tall spears, stood in the sunlight under fantastic head-dresses of spotted skins, warlike and still in statuesque repose. And from right to left along the lighted shore moved a wild and gorgeous apparition of a woman.

"She walked with measured steps, draped in striped and

fringed cloths, treading the earth proudly, with a slight jingle and flash of barbarous ornaments. She carried her head high; her hair was done in the shape of a helmet; she had brass leggings to the knees, brass wire gauntlets to the elbows, a crimson spot on her tawny cheek, innumerable necklaces of glass beads on her neck; bizarre things, charms, gifts of witch-men, that hung about her, glittered and trembled at every step. She must have had the value of several elephant tusks upon her. She was savage and superb, wild-eyed and magnificent; there was something ominous and stately in her deliberate progress. And in the hush that had fallen suddenly upon the whole sorrowful land, the immense wilderness, the colossal body of the fecund and mysterious life seemed to look at her, pensive, as though it had been looking at the image of its own tenebrous and passionate soul.

"She came abreast of the steamer, stood still, and faced us. Her long shadow fell to the water's edge. Her face had a tragic and fierce aspect of wild sorrow and of dumb pain mingled with the fear of some struggling, half-shaped resolve. She stood looking at us without a stir, and like the wilderness itself, with an air of brooding over an inscrutable purpose. A whole minute passed, and then she made a step forward. There was a low jingle, a glint of yellow metal, a sway of fringed draperies, and she stopped as if her heart had failed her. The young fellow by my side growled. The pilgrims murmured at my back. She looked at us all as if her life had depended upon the unswerving steadiness of her glance. Suddenly she opened her bared arms and threw them up rigid above her head, as though in an uncontrollable desire to touch the sky, and at the same time the swift shadows darted out on the earth, swept around on the river, gathering the steamer into a shadowy embrace. A formidable silence hung over the scene.

"She turned away slowly, walked on, following the bank, and passed into the bushes to the left. Once only her eyes gleamed back at us in the dusk of the thickets before she disappeared.

" 'If she had offered to come aboard I really think I would have tried to shoot her,' said the man of patches, nervously. 'I had been risking my life every day for the last fortnight to keep her out of the house. She got in one day and kicked up a row about those miserable rags I picked up in the storeroom to mend my clothes with. I wasn't decent. At least it must have been that, for she talked like a fury to Kurtz for an hour, pointing at me now and then. I don't understand the dialect of this tribe. Luckily for me, I fancy Kurtz felt too ill that day to care, or there would have been mischief. I don't understand . . . No— it's too much for me. Ah, well, it's all over now.'

"At this moment I heard Kurtz's deep voice behind the curtain: 'Save me!—save the ivory, you mean. Don't tell me. Save *me!* Why, I've had to save you. You are interrupting my plans now. Sick! Sick! Not so sick as you would like to believe. Never mind. I'll carry my ideas out yet—I will return. I'll show you what can be done. You with your little peddling notions —you are interfering with me. I will return. I . . .'

"The manager came out. He did me the honour to take me under the arm and lead me aside. 'He is very low, very low,' he said. He considered it necessary to sigh, but neglected to be consistently sorrowful. 'We have done all we could for him— haven't we? But there is no disguising the fact, Mr. Kurtz has done more harm than good to the Company. He did not see the time was not ripe for vigorous action. Cautiously, cautiously—that's my principle. We must be cautious yet. The district is closed to us for a time. Deplorable! Upon the whole, the trade will suffer. I don't deny there is a remarkable quantity of ivory—mostly fossil. We must save it, at all events—but look how precarious the position is—and why? Because the method is unsound.' 'Do you,' said I, looking at the shore, 'call it "unsound method"?' 'Without doubt,' he exclaimed, hotly. 'Don't you?' . . .

" 'No method at all,' I murmured after a while. 'Exactly,'

he exulted. 'I anticipated this. Shows a complete want of judgement. It is my duty to point it out in the proper quarter.' 'Oh,'
said I, 'that fellow—what's his name?—the brickmaker, will
make a readable report for you.' He appeared confounded for
a moment. It seemed to me I had never breathed an atmosphere
so vile, and I turned mentally to Kurtz for relief—positively for
relief. 'Nevertheless I think Mr. Kurtz is a remarkable man,' I
said with emphasis. He started, dropped on me a cold heavy
glance, said very quietly, 'He *was*,' and turned his back on me.
My hour of favour was over; I found myself lumped along with
Kurtz as a partisan of methods for which the time was not ripe:
I was unsound! Ah! but it was something to have at least a
choice of nightmares.

"I had turned to the wilderness really, not to Mr. Kurtz, who,
I was ready to admit, was as good as buried. And for a moment
it seemed to me as if I also were buried in a vast grave full of
unspeakable secrets. I felt an intolerable weight oppressing my
breast, the smell of the damp earth, the unseen presence of
victorious corruption, the darkness of an impenetrable night . . .
The Russian tapped me on the shoulder. I heard him mumbling
and stammering something about 'brother seaman—couldn't
conceal—knowledge of matters that would affect Mr. Kurtz's
reputation.' I waited. For him evidently Mr. Kurtz was not in
his grave; I suspect that for him Mr. Kurtz was one of the
immortals. 'Well!' said I at last, 'speak out. As it happens, I
am Mr. Kurtz's friend—in a way.'

"He stated with a good deal of formality that had we not
been 'of the same profession,' he would have kept the matter
to himself without regard to consequences. 'He suspected there
was an active ill will towards him on the part of these white
men that—' 'You are right,' I said, remembering a certain
conversation I had overheard. 'The manager thinks you ought
to be hanged.' He showed a concern at this intelligence which
amused me at first. 'I had better get out of the way quietly,'

he said, earnestly. 'I can do no more for Kurtz now, and they
would soon find some excuse. What's to stop them? There's a
military post three hundred miles from here.' 'Well, upon my
word,' said I, 'perhaps you had better go if you have any friends
amongst the savages near by.' 'Plenty,' he said. 'They are simple
people—and I want nothing, you know.' He stood biting his
lip, then: 'I don't want any harm to happen to these whites
here, but of course I was thinking of Mr. Kurtz's reputation—
but you are a brother seaman and—' 'All right,' said I, after a
time. 'Mr. Kurtz's reputation is safe with me.' I did not know
how truly I spoke.

"He informed me, lowering his voice, that it was Kurtz who
had ordered the attack to be made on the steamer. 'He hated
sometimes the idea of being taken away—and then again . . .
But I don't understand these matters. I am a simple man. He
thought it would scare you away—that you would give it up,
thinking him dead. I could not stop him. Oh, I had an awful
time of it this last month.' 'Very well,' I said. 'He is all right
now.' 'Ye-e-es,' he muttered, not very convinced apparently.
'Thanks,' said I; 'I shall keep my eyes open.' 'But quiet—eh?'
he urged, anxiously. 'It would be awful for his reputation if
anybody here—' I promised a complete discretion with great
gravity. 'I have a canoe and three black fellows waiting not very
far. I am off. Could you give me a few Martini-Henry cartridges?'
I could, and did, with proper secrecy. He helped himself, with
a wink at me, to a handful of my tobacco. 'Between sailors—
you know—good English tobacco.' At the door of the pilot-
house he turned round—'I say, haven't you a pair of shoes you
could spare?' He raised one leg. 'Look.' The soles were tied with
knotted strings sandal-wise under his bare feet. I rooted out an
old pair, at which he looked with admiration before tucking it
under his left arm. One of his pockets (bright red) was bulging
with cartridges, from the other (dark blue) peeped 'Towson's
Inquiry,' etc., etc. He seemed to think himself excellently well

equipped for a renewed encounter with the wilderness. 'Ah! I'll never, never meet such a man again. You ought to have heard him recite poetry—his own, too, it was, he told me. Poetry!' He rolled his eyes at the recollection of these delights. 'Oh, he enlarged my mind!' 'Good-bye,' said I. He shook hands and vanished in the night. Sometimes I ask myself whether I had ever really seen him—whether it was possible to meet such a phenomenon! . . .

"When I woke up shortly after midnight his warning came to my mind with its hint of danger that seemed, in the starred darkness, real enough to make me get up for the purpose of having a look round. On the hill a big fire burned, illuminating fitfully a crooked corner of the station-house. One of the agents with a picket of a few of our blacks, armed for the purpose, was keeping guard over the ivory; but deep within the forest, red gleams that wavered, that seemed to sink and rise from the ground amongst confused columnar shapes of intense blackness, showed the exact position of the camp where Mr. Kurtz's adorers were keeping their uneasy vigil. The monotonous beating of a big drum filled the air with muffled shocks and a lingering vibration. A steady droning sound of many men chanting each to himself some weird incantation came out from the black, flat wall of the woods as the humming of bees comes out of a hive, and had a strange narcotic effect upon my half-awake senses. I believe I dozed off leaning over the rail, till an abrupt burst of yells, an overwhelming outbreak of a pent-up and mysterious frenzy, woke me up in a bewildered wonder. It was cut short all at once, and the low droning went on with an effect of audible and soothing silence. I glanced casually into the little cabin. A light was burning within, but Mr. Kurtz was not there.

"I think I would have raised an outcry if I had believed my eyes. But I didn't believe them at first—the thing seemed so impossible. The fact is I was completely unnerved by a sheer blank fright, pure abstract terror, unconnected with any distinct shape of physical danger. What made this emotion so overpowering was—how shall I define it?—the moral shock I received, as if something altogether monstrous, intolerable to thought and odious to the soul, had been thrust upon me unexpectedly. This lasted of course the merest fraction of a second, and then the usual sense of commonplace, deadly danger, the possibility of a sudden onslaught and massacre, or something of the kind, which I saw impending, was positively welcome and composing. It pacified me, in fact, so much, that I did not raise an alarm.

"There was an agent buttoned up inside an ulster and sleeping on a chair on deck within three feet of me. The yells had not awakened him; he snored very slightly; I left him to his slumbers and leaped ashore. I did not betray Mr. Kurtz—it was ordered I should never betray him—it was written I should be loyal to the nightmare of my choice. I was anxious to deal with this shadow by myself alone—and to this day I don't know why I was so jealous of sharing with any one the peculiar blackness of that experience.

"As soon as I got on the bank I saw a trail—a broad trail through the grass. I remember the exultation with which I said to myself, 'He can't walk—he is crawling on all-fours—I've got him.' The grass was wet with dew. I strode rapidly with clenched fists. I fancy I had some vague notion of falling upon him and giving him a drubbing. I don't know. I had some imbecile thoughts. The knitting old woman with the cat obtruded herself upon my memory as a most improper person to be sitting at the other end of such an affair. I saw a row of pilgrims squirting lead in the air out of Winchesters held to the hip. I thought I would never get back to the steamer, and imagined myself living alone and unarmed in the woods to an advanced age. Such silly things—you know. And I remember I confounded the beat of the drum with the beating of my heart, and was pleased at its calm regularity.

"I kept to the track though—then stopped to listen. The night was very clear; a dark blue space, sparkling with dew and starlight, in which black things stood very still. I thought I could see a kind of motion ahead of me. I was strangely cocksure of everything that night. I actually left the track and ran in a wide semicircle (I verily believe chuckling to myself) so as to get in front of that stir, of that motion I had seen—if indeed I had seen anything. I was circumventing Kurtz as though it had been a boyish game.

"I came upon him, and, if he had not heard me coming, I would have fallen over him, too, but he got up in time. He rose, unsteady, long, pale, indistinct, like a vapour exhaled by the earth, and swayed slightly, misty and silent before me; while at my back the fires loomed between the trees, and the murmur of many voices issued from the forest. I had cut him off cleverly; but when actually confronting him I seemed to come to my senses, I saw the danger in its right proportion. It was by no means over yet. Suppose he began to shout? Though he could hardly stand, there was still plenty of vigour in his voice. 'Go away—hide yourself,' he said, in that profound tone. It was very awful. I glanced back. We were within thirty yards from the nearest fire. A black figure stood up, strode on long black legs, waving long black arms, across the glow. It had horns—antelope horns, I think—on its head. Some sorcerer, some witchman, no doubt: it looked fiend-like enough. 'Do you know what you are doing?' I whispered. 'Perfectly,' he answered, raising his voice for that single word: it sounded to me far off and yet loud, like a hail through a speaking-trumpet. If he makes a row we are lost, I thought to myself. This clearly was not a case for fisticuffs, even apart from the very natural aversion I had to beat that Shadow—this wandering and tormented thing. 'You will be lost,' I said—'utterly lost.' One gets sometimes such a flash of inspiration, you know. I did say the right thing, though indeed he could not have been more irretrievably lost than he was at this very moment, when the foundations of our intimacy were being laid—to endure—to endure—even to the end—even beyond.

" 'I had immense plans,' he muttered irresolutely. 'Yes,' said I; 'but if you try to shout I'll smash your head with—' There was not a stick or a stone near. 'I will throttle you for good,' I corrected myself. 'I was on the threshold of great things,' he pleaded, in a voice of longing, with a wistfulness of tone that made my blood run cold. 'And now for this stupid scoundrel—' 'Your success in Europe is assured in any case,' I affirmed, steadily. I did not want to have the throttling of him, you understand—and indeed it would have been very little use for any practical purpose. I tried to break the spell—the heavy, mute spell of the wilderness—that seemed to draw him to its pitiless breast by the awakening of forgotten and brutal instincts, by the memory of gratified and monstrous passions. This alone, I was convinced, had driven him out to the edge of the forest, to the bush, towards the gleam of fires, the throb of drums, the drone of weird incantations; this alone had beguiled his unlawful soul beyond the bounds of permitted aspirations. And, don't you see, the terror of the position was not in being knocked on the head—though I had a very lively sense of that danger, too—but in this, that I had to deal with a being to whom I could not appeal in the name of anything high or low. I had, even like the niggers, to invoke him—himself—his own exalted and incredible degradation. There was nothing either above or below him, and I knew it. He had kicked himself loose of the earth. Confound the man! he had kicked the very earth to pieces. He was alone, and I before him did not know whether I stood on the ground or floated in the air. I've been telling you what we said—repeating the phrases we pronounced—but what's the good? They were common everyday words—the familiar, vague sounds exchanged on every waking day of life. But what of that? They had behind them, to my mind, the terrific suggestiveness of words heard in dreams, of phrases spoken in night-mares. Soul! If anybody had ever struggled with a soul, I am the man. And I wasn't arguing with a lunatic either. Believe me or not, his intelligence was perfectly clear—concentrated, it is true, upon himself with horrible intensity, yet clear; and therein was my only chance—barring, of course, the killing him there and then, which wasn't so good, on account of unavoidable noise. But his soul was mad. Being alone in the wilderness, it had looked within itself, and, by heavens! I tell you, it had gone mad. I had—for my sins, I suppose—to go through the ordeal of looking into it myself. No eloquence could have been so withering to one's belief in mankind as his final burst of sincerity. He struggled with himself, too. I saw it—I heard it. I saw the inconceivable mystery of a soul that knew no restraint, no faith, and no fear, yet struggling blindly with itself. I kept my head pretty well; but when I had him at last stretched on the couch, I wiped my forehead, while my legs shook under me as though I had carried half a ton on my back down that hill. And yet I had only supported him, his bony arm clasped round my neck—and he was not much heavier than a child.

"When next day we left at noon, the crowd, of whose presence behind the curtain of trees I had been acutely conscious all the time, flowed out of the woods again, filled the clearing, covered the slope with a mass of naked, breathing, quivering, bronze bodies. I steamed up a bit, then swung downstream, and two thousand eyes followed the evolutions of the splashing, thumping, fierce river-demon beating the water with its terrible tail and breathing black smoke into the air. In front of the first rank, along the river, three men, plastered with bright red earth from head to foot, strutted to and fro restlessly. When we came abreast again, they faced the river, stamped their feet, nodded their horned heads, swayed their scarlet bodies; they shook towards the fierce river-demon a bunch of black feathers, a mangy skin with a pendent tail—something that looked like a dried gourd; they shouted periodically together strings of amaz-ing words that resembled no sounds of human language; and the deep murmurs of the crowd, interrupted suddenly, were like the responses of some satanic litany.

"We had carried Kurtz into the pilot-house: there was more air there. Lying on the couch, he stared through the open shutter. There was an eddy in the mass of human bodies, and the woman with the helmeted head and tawny cheeks rushed out to the very brink of the stream. She put out her hands, shouted some-thing, and all that wild mob took up the shout in a roaring chorus of articulated, rapid, breathless utterance.

" 'Do you understand this?' I asked.

"He kept on looking out past me with fiery, longing eyes, with a mingled expression of wistfulness and hate. He made no answer, but I saw a smile, a smile of indefinable meaning, appear on his colourless lips that a moment after twitched convulsively. 'Do I not?' he said slowly, gasping, as if the words had been torn out of him by a supernatural power.

"I pulled the string of the whistle, and I did this because I saw the pilgrims on deck getting out their rifles with an air of anticipating a jolly lark. At the sudden screech there was a movement of abject terror through that wedged mass of bodies. 'Don't! don't you frighten them away,' cried someone on deck disconsolately. I pulled the string time after time. They broke and ran, they leaped, they crouched, they swerved, they dodged the flying terror of the sound. The three red chaps had fallen flat, face down on the shore, as though they had been shot dead. Only the barbarous and superb woman did not so much as flinch and stretched tragically her bare arms after us over the sombre and glittering river.

"And then that imbecile crowd down on the deck started their little fun, and I could see nothing more for smoke.

"The brown current ran swiftly out of the heart of darkness, bearing us down towards the sea with twice the speed of our

upward progress; and Kurtz's life was running swiftly, too, ebbing, ebbing out of his heart into the sea of inexorable time.
The manager was very placid, he had no vital anxieties now,
he took us both in with a comprehensive and satisfied glance:
the 'affair' had come off as well as could be wished. I saw the
time approaching when I would be left alone of the party of
'unsound method.' The pilgrims looked upon me with disfa-
vour. I was, so to speak, numbered with the dead. It is strange
how I accepted this unforeseen partnership, this choice of night-
mares forced upon me in the tenebrous land invaded by these
mean and greedy phantoms.

"Kurtz discoursed. A voice! a voice! It rang deep to the very
last. It survived his strength to hide in the magnificent folds of
eloquence the barren darkness of his heart. Oh, he struggled!
he struggled! The wastes of his weary brain were haunted by
shadowy images now—images of wealth and fame revolving
obsequiously round his unextinguishable gift of noble and lofty
expression. My Intended, my station, my career, my ideas—
these were the subjects for the occasional utterances of elevated
sentiments. The shade of the original Kurtz frequented the bed-
side of the hollow sham, whose fate it was to be buried presently
in the mould of primeval earth. But both the diabolic love and
the unearthly hate of the mysteries it had penetrated fought for
the possession of that soul satiated with primitive emotions,
avid of lying fame, of sham distinction, of all the appearances
of success and power.

"Sometimes he was contemptibly childish. He desired to have
kings meet him at railway-stations on his return from some
ghastly Nowhere, where he intended to accomplish great things.
'You show them you have in you something that is really profitable, and then there will be no limits to the recognition of
your ability,' he would say. 'Of course you must take care of
the motives—right motives—always.' The long reaches that were
like one and the same reach, monotonous bends that were exactly

alike, slipped past the steamer with their multitude of secular trees looking patiently after this grimy fragment of another world, the forerunner of change, of conquest, of trade, of massacres, of blessings. I looked ahead—piloting. 'Close the shutter,' said Kurtz suddenly one day; 'I can't bear to look at this.' I did so. There was a silence. 'Oh, but I will wring your heart yet!' he cried at the invisible wilderness.

"We broke down—as I had expected—and had to lie up for repairs at the head of an island. This delay was the first thing that shook Kurtz's confidence. One morning he gave me a packet of papers and a photograph—the lot tied together with a shoe-string. 'Keep this for me,' he said. 'This noxious fool' (meaning the manager) 'is capable of prying into my boxes when I am not looking.' In the afternoon I saw him. He was lying on his back with closed eyes, and I withdrew quietly, but I heard him mutter, 'Live rightly, die, die . . .' I listened. There was nothing more. Was he rehearsing some speech in his sleep, or was it a fragment of a phrase from some newspaper article? He had been writing for the papers and meant to do so again, 'for the furthering of my ideas. It's a duty.'

"His was an impenetrable darkness. I looked at him as you peer down at a man who is lying at the bottom of a precipice where the sun never shines. But I had not much time to give him, because I was helping the engine-driver to take to pieces the leaky cylinders, to straighten a bent connecting-rod, and in other such matters. I lived in an infernal mess of rust, filings, nuts, bolts, spanners, hammers, ratchet-drills—things I abominate, because I don't get on with them. I tended the little forge we fortunately had aboard; I toiled wearily in a wretched scrap-heap—unless I had the shakes too bad to stand.

"One evening coming in with a candle I was startled to hear him say a little tremulously, 'I am lying here in the dark waiting for death.' The light was within a foot of his eyes. I forced myself to murmur, 'Oh, nonsense!' and stood over him as if transfixed.

"Anything approaching the change that came over his features I have never seen before, and hope never to see again. Oh, I wasn't touched. I was fascinated. It was as though a veil had been rent. I saw on that ivory face the expression of sombre pride, of ruthless power, of craven terror—of an intense and hopeless despair. Did he live his life again in every detail of desire, temptation, and surrender during that supreme moment of complete knowledge? He cried in a whisper at some image, at some vision—he cried out twice, a cry that was no more than a breath—

" 'The horror! The horror!'

"I blew the candle out and left the cabin. The pilgrims were dining in the mess-room, and I took my place opposite the manager, who lifted his eyes to give me a questioning glance, which I successfully ignored. He leaned back, serene, with that peculiar smile of his sealing the unexpressed depths of his meanness. A continuous shower of small flies streamed upon the lamp, upon the cloth, upon our hands and faces. Suddenly the manager's boy put his insolent black head in the doorway, and said in a tone of scathing contempt—

" 'Mistah Kurtz—he dead.'

"All the pilgrims rushed out to see. I remained, and went on with my dinner. I believe I was considered brutally callous. However, I did not eat much. There was a lamp in there—light, don't you know—and outside it was so beastly, beastly dark. I went no more near the remarkable man who had pronounced a judgement upon the adventures of his soul on this earth. The voice was gone. What else had been there? But I am of course aware that next day the pilgrims buried something in a muddy hole.

"And then they very nearly buried me.

"However, as you see, I did not go to join Kurtz there and then. I did not. I remained to dream the nightmare out to the end, and to show my loyalty to Kurtz once more. Destiny. My

Destiny! Droll thing life is—that mysterious arrangement of merciless logic for a futile purpose. The most you can hope from it is some knowledge of yourself—that comes too late—a crop of unextinguishable regrets. I have wrestled with death. It is the most unexciting contest you can imagine. It takes place in an impalpable greyness, with nothing underfoot, with nothing around, without spectators, without clamour, without glory, without the great desire of victory, without the great fear of defeat, in a sickly atmosphere of tepid scepticism, without much belief in your own right, and still less in that of your adversary. If such is the form of ultimate wisdom, then life is a greater riddle than some of us think it to be. I was within a hair's-breadth of the last opportunity for pronouncement, and I found with humiliation that probably I would have nothing to say. This is the reason why I affirm that Kurtz was a remarkable man. He had something to say. He said it. Since I had peeped over the edge myself, I understand better the meaning of his stare, that could not see the flame of the candle, but was wide enough to embrace the whole universe, piercing enough to penetrate all the hearts that beat in the darkness. He had summed up—he had judged. 'The horror!' He was a remarkable man. After all, this was the expression of some sort of belief; it had candour, it had conviction, it had a vibrating note of revolt in its whisper, it had the appalling face of a glimpsed truth—the strange commingling of desire and hate. And it is not my own extremity I remember best—a vision of greyness without form filled with physical pain, and a careless contempt for the evanescence of all things—even of this pain itself. No! It is his extremity that I seem to have lived through. True, he had made that last stride, he had stepped over the edge, while I had been permitted to draw back my hesitating foot. And perhaps in this is the whole difference; perhaps all the wisdom, and all truth, and all sincerity, are just compressed into that inappreciable moment of time in which we step over the threshold of the invisible. Perhaps! I like to think my summing-up would not have been a word of careless contempt. Better his cry—much better. It was an affirmation, a moral victory, paid for by innumerable defeats, by abominable terrors, by abominable satisfactions. But it was a victory! That is why I have remained loyal to Kurtz to the last, and even beyond, when a long time after I heard once more, not his own voice, but the echo of his magnificent eloquence thrown to me from a soul as translucently pure as a cliff of crystal.

"No, they did not bury me, though there is a period of time which I remember mistily, with a shuddering wonder, like a passage through some inconceivable world that had no hope in it and no desire. I found myself back in the sepulchral city resenting the sight of people hurrying through the streets to filch a little money from each other, to devour their infamous cookery, to gulp their unwholesome beer, to dream their insignificant and silly dreams. They trespassed upon my thoughts. They were intruders whose knowledge of life was to me an irritating pretence, because I felt so sure they could not possibly know the things I knew. Their bearing, which was simply the bearing of commonplace individuals going about their business in the assurance of perfect safety, was offensive to me like the outrageous flauntings of folly in the face of a danger it is unable to comprehend. I had no particular desire to enlighten them, but I had some difficulty in restraining myself from laughing in their faces, so full of stupid importance. I daresay I was not very well at that time. I tottered about the streets—there were various affairs to settle—grinning bitterly at perfectly respectable persons. I admit my behaviour was inexcusable, but then my temperature was seldom normal in these days. My dear aunt's endeavours to 'nurse up my strength' seemed altogether beside the mark. It was not my strength that wanted nursing, it was my imagination that wanted soothing. I kept the bundle of papers given me by Kurtz, not knowing exactly what to do with it. His mother had died lately, watched over, as I was told, by his Intended. A clean-shaven man, with an official manner and wearing gold-rimmed spectacles, called on me one day and made inquiries, at first circuitous, afterwards suavely pressing, about what he was pleased to denominate certain 'documents.' I was not surprised, because I had had two rows with the manager on the subject out there. I had refused to give up the smallest scrap out of that package, and I took the same attitude with the spectacled man. He became darkly menacing at the last, and with much heat argued that the Company had the right to every bit of information about its 'territories.' And said he, 'Mr. Kurtz's knowledge of unexplored regions must have been necessarily extensive and peculiar—owing to his great abilities and to the deplorable circumstances in which he had been placed: therefore—' I assured him Mr. Kurtz's knowledge, however extensive, did not bear upon the problems of commerce or administration. He invoked then the name of science. 'It would be an incalculable loss if,' etc., etc. I offered him the report on the 'Suppression of Savage Customs,' with the postscriptum torn off. He took it up eagerly, but ended by sniffing at it with an air of contempt. 'This is not what we had a right to expect,' he remarked. 'Expect nothing else,' I said. 'There are only private letters.' He withdrew upon some threat of legal proceedings, and I saw him no more; but another fellow, calling himself Kurtz's cousin, appeared two days later, and was anxious to hear all the details about his dear relative's last moments. Incidentally he gave me to understand that Kurtz had been essentially a great musician. 'There was the making of an immense success,' said the man, who was an organist, I believe, with lank grey hair flowing over a greasy coat-collar. I had no reason to doubt his statement; and to this day I am unable to say what was Kurtz's profession, whether he ever had any—which was the greatest of his talents. I had taken him for a painter who wrote for the papers, or else for a journalist who could paint—

but even the cousin (who took snuff during the interview) could not tell me what he had been—exactly. He was a universal genius—on that point I agreed with the old chap, who thereupon blew his nose noisily into a large cotton handkerchief and withdrew in senile agitation, bearing off some family letters and memoranda without importance. Ultimately a journalist anxious to know something of the fate of his 'dear colleague' turned up. This visitor informed me Kurtz's proper sphere ought to have been politics 'on the popular side.' He had furry straight eyebrows, bristly hair cropped short, an eye-glass on a broad ribbon, and, becoming expansive, confessed his opinion that Kurtz really couldn't write a bit—'but heavens! how that man could talk. He electrified large meetings. He had faith—don't you see?— he had the faith. He could get himself to believe anything— anything. He would have been a splendid leader of an extreme party.' 'What party?' I asked. 'Any party,' answered the other. 'He was an—an—extremist.' Did I not think so? I assented. Did I know, he asked, with a sudden flash of curiosity, 'what it was that had induced him to go out there?' 'Yes,' said I, and forthwith handed him the famous Report for publication, if he thought fit. He glanced through it hurriedly, mumbling all the time, judged 'it would do,' and took himself off with this plunder.

"Thus I was left at last with a slim packet of letters and the girl's portrait. She struck me as beautiful—I mean she had a beautiful expression. I know that the sunlight can be made to lie, too, yet one felt that no manipulation of light and pose could have conveyed the delicate shade of truthfulness upon those features. She seemed ready to listen without mental reservation, without suspicion, without a thought for herself. I concluded I would go and give her back her portrait and those letters myself. Curiosity? Yes; and also some other feelings perhaps. All that had been Kurtz's had passed out of my hands: his soul, his body, his station, his plans, his ivory, his career.

There remained only his memory and his Intended—and I wanted to give that up, too, to the past, in a way—to surrender personally all that remained of him with me to that oblivion which is the last word of our common fate. I don't defend myself. I had no clear perception of what it was I really wanted. Perhaps it was an impulse of unconscious loyalty, or the fulfilment of one of these ironic necessities that lurk in the facts of human existence. I don't know. I can't tell. But I went.

"I thought his memory was like the other memories of the dead that accumulate in every man's life—a vague impress on the brain of shadows that had fallen on it in their swift and final passage; but before the high and ponderous door, between the tall houses of a street as still and decorous as a well-kept alley in a cemetery, I had a vision of him on the stretcher, opening his mouth voraciously, as if to devour all the earth with all its mankind. He lived then before me; he lived as much as he had ever lived—a shadow insatiable of splendid appearances, of frightful realities; a shadow darker than the shadow of the night, and draped nobly in the folds of a gorgeous eloquence. The vision seemed to enter the house with me—the stretcher, the phantom-bearers, the wild crowd of obedient worshippers, the gloom of the forests, the glitter of the reach between the murky bends, the beat of the drum, regular and muffled like the beating of a heart—the heart of a conquering darkness. It was a moment of triumph for the wilderness, an invading and vengeful rush which, it seemed to me, I would have to keep back alone for the salvation of another soul. And the memory of what I had heard him say afar there, with the horned shapes stirring at my back, in the glow of fires, within the patient woods, those broken phrases came back to me, were heard again in their ominous and terrifying simplicity. I remembered his abject pleading, his abject threats, the colossal scale of his vile desires, the meanness, the torment, the tempestuous anguish of his soul. And later on I seemed to see his collected languid manner, when he said one day, 'This lot of ivory now is really mine. The Company did not pay for it. I collected it myself at a very great personal risk. I am afraid they will try to claim it as theirs though. H'm. It is a difficult case. What do you think I ought to do—resist? Eh? I want no more than justice.' . . . He wanted no more than justice—no more than justice. I rang the bell before a mahogany door on the first floor, and while I waited he seemed to stare at me out of the glassy panel—stare with that wide and immense stare embracing, condemning, loathing all the universe. I seemed to hear the whispered cry, 'The horror! The horror!'

"The dusk was falling. I had to wait in a lofty drawing-room with three long windows from floor to ceiling that were like three luminous and bedraped columns. The bent gilt legs and backs of the furniture shone in indistinct curves. The tall marble fireplace had a cold and monumental whiteness. A grand piano stood massively in a corner; with dark gleams on the flat surfaces like a sombre and polished sarcophagus. A high door opened—closed. I rose.

"She came forward, all in black, with a pale head, floating towards me in the dusk. She was in mourning. It was more than a year since his death, more than a year since the news came; she seemed as though she would remember and mourn for ever. She took both my hands in hers and murmured, 'I had heard you were coming.' I noticed she was not very young—I mean not girlish. She had a mature capacity for fidelity, for belief, for suffering. The room seemed to have grown darker, as if all the sad light of the cloudy evening had taken refuge on her forehead. This fair hair, this pale visage, this pure brow, seemed surrounded by an ashy halo from which the dark eyes looked out at me. Their glance was guileless, profound, confident, and trustful. She carried her sorrowful head as though she were proud of that sorrow, as though she would say, I—I alone know how to mourn for him as he deserves. But while

we were still shaking hands, such a look of awful desolation
came upon her face that I perceived she was one of those creatures
that are not the playthings of Time. For her he had died only
yesterday. And, by Jove! the impression was so powerful that
for me, too, he seemed to have died only yesterday—nay, this
very minute. I saw her and him in the same instant of time—
his death and her sorrow—I saw her sorrow in the very moment
of his death. Do you understand? I saw them together—I heard
them together. She had said, with a deep catch of the breath,
'I have survived' while my strained ears seemed to hear distinctly, mingled with her tone of despairing regret, the summing
up whisper of his eternal condemnation. I asked myself what I
was doing there, with a sensation of panic in my heart as though
I had blundered into a place of cruel and absurd mysteries not
fit for a human being to behold. She motioned me to a chair.
We sat down. I laid the packet gently on the little table, and
she put her hand over it . . . 'You knew him well,' she murmured, after a moment of mourning silence.

" 'Intimacy grows quickly out there,' I said. 'I knew him as
well as it is possible for one man to know another.'

" 'And you admired him,' she said. 'It was impossible to
know him and not to admire him. Was it?'

" 'He was a remarkable man,' I said, unsteadily. Then before
the appealing fixity of her gaze, that seemed to watch for more
words on my lips, I went on. 'It was impossible not to—'

" 'Love him,' she finished eagerly, silencing me into an appalled dumbness. 'How true! how true! But when you think
that no one knew him so well as I! I had all his noble confidence.
I knew him best.'

" 'You knew him best,' I repeated. And perhaps she did.
But with every word spoken the room was growing darker, and
only her forehead, smooth and white, remained illumined by
the unextinguishable light of belief and love.

" 'You were his friend,' she went on. 'His friend,' she re-peated, a little louder. 'You must have been, if he had given you this, and sent you to me. I feel I can speak to you—and oh! I must speak. I want you—you who have heard his last words—to know I have been worthy of him . . . It is not pride . . . Yes! I am proud to know I understood him better than any one on earth—he told me so himself. And since his mother died I have had no one—no one—to—to—'

"I listened. The darkness deepened. I was not even sure whether he had given me the right bundle. I rather suspect he wanted me to take care of another batch of his papers which, after his death, I saw the manager examining under the lamp. And the girl talked, easing her pain in the certitude of my sympathy; she talked as thirsty men drink. I had heard that her engagement with Kurtz had been disapproved by her people. He wasn't rich enough or something. And indeed I don't know whether he had not been a pauper all his life. He had given me some reason to infer that it was his impatience of comparative poverty that drove him out there.

" ' . . . Who was not his friend who had heard him speak once?' she was saying. 'He drew men towards him by what was best in them.' She looked at me with intensity. 'It is the gift of the great,' she went on, and the sound of her low voice seemed to have the accompaniment of all the other sounds, full of mystery, desolation, and sorrow, I had ever heard—the ripple of the river, the soughing of the trees swayed by the wind, the murmurs of the crowds, the faint ring of incomprehensible words cried from afar, the whisper of a voice speaking from beyond the threshold of an eternal darkness. 'But you have heard him! You know!' she cried.

" 'Yes, I know,' I said with something like despair in my heart, but bowing my head before the faith that was in her, before that great and saving illusion that shone with an unearthly glow in the darkness, in the triumphant darkness from which

I could not have defended her—from which I could not even defend myself.

" 'What a loss to me—to us!'—she corrected herself with beautiful generosity; then added in a murmur, 'To the world.' By the last gleams of twilight I could see the glitter of her eyes, full of tears—of tears that would not fall.

" 'I have been very happy—very fortunate—very proud,' she went on. 'Too fortunate. Too happy for a little while. And now I am unhappy for—for life.'

"She stood up; her fair hair seemed to catch all the remaining light in a glimmer of gold. I rose, too.

" 'And of all this,' she went on, mournfully, 'of all his promise, and of all his greatness, of his generous mind, of his noble heart, nothing remains—nothing but a memory. You and I—'

" 'We shall always remember him,' I said hastily.

" 'No!' she cried. 'It is impossible that all this should be lost—that such a life should be sacrificed to leave nothing—but sorrow. You know what vast plans he had. I knew of them, too—I could not perhaps understand—but others knew of them. Something must remain. His words, at least, have not died.'

" 'His words will remain,' I said.

" 'And his example,' she whispered to herself. 'Men looked up to him—his goodness shone in every act. His example—'

" 'True,' I said; 'his example, too. Yes, his example. I forgot that.'

" 'But I do not. I cannot—I cannot believe—not yet. I cannot believe that I shall never see him again, that nobody will see him again, never, never, never.'

"She put out her arms as if after a retreating figure, stretching them back and with clasped pale hands across the fading and narrow sheen of the window. Never see him! I saw him clearly enough then. I shall see this eloquent phantom as long as I live, and I shall see her, too, a tragic and familiar Shade, resembling in this gesture another one, tragic also, and bedecked with powerless charms, stretching bare brown arms over the glitter of the infernal stream, the stream of darkness. She said suddenly very low, 'He died as he lived.'

" 'His end,' said I, with dull anger stirring in me, 'was in every way worthy of his life.'

" 'And I was not with him,' she murmured. My anger subsided before a feeling of infinite pity.

" 'Everything that could be done—' I mumbled.

" 'Ah, but I believed in him more than any one on earth—more than his own mother, more than—himself. He needed me! Me! I would have treasured every sigh, every word, every sign, every glance.'

"I felt like a chill grip on my chest. 'Don't,' I said, in a muffled voice.

" 'Forgive me. I—I—have mourned so long in silence—in silence . . . You were with him—to the last? I think of his loneliness. Nobody near to understand him as I would have understood. Perhaps no one to hear . . . '

" 'To the very end,' I said, shakily. 'I heard his very last words . . .' I stopped in fright.

" 'Repeat them,' she murmured in a heart-broken tone. 'I want—I want—something—something—to—to live with.'

"I was on the point of crying at her, 'Don't you hear them?' The dusk was repeating them in a persistent whisper all around us, in a whisper that seemed to swell menacingly like the first whisper of a rising wind. 'The horror! The horror!'

" 'His last word—to live with,' she insisted. 'Don't you understand I loved him—I loved him—I loved him!'

"I pulled myself together and spoke slowly.

" 'The last word he pronounced was—your name.'

"I heard a light sigh and then my heart stood still, stopped dead short by an exulting and terrible cry, by the cry of inconceivable triumph and of unspeakable pain. 'I knew it—I was sure!' . . . She knew. She was sure. I heard her weeping; she had hidden her face in her hands. It seemed to me that the house would collapse before I could escape, that the heavens would fall upon my head. But nothing happened. The heavens do not fall for such a trifle. Would they have fallen, I wonder, if I had rendered Kurtz that justice which was his due? Hadn't he said he wanted only justice? But I couldn't. I could not tell her. It would have been too dark—too dark altogether. . . ."

Marlow ceased, and sat apart, indistinct and silent, in the pose of a meditating Buddha. Nobody moved for a time. "We have lost the first of the ebb," said the Director, suddenly. I raised my head. The offing was barred by a black bank of clouds, and the tranquil waterway leading to the uttermost ends of the earth flowed sombre under an overcast sky—seemed to lead into the heart of an immense darkness.

IMMANUEL KANT, the son of a saddler, was born in 1724 in Konigsberg, East Prussia. Kant studied science and philosophy at the University of Konigsberg. He completed his degree and was appointed a lecturer there in 1755. Kant taught and wrote about a broad range of subjects, including metaphysics, logic, ethics, geography, anthropology, mathematics, physics, astronomy, geology, meteorology, and fireworks. "His lectures," a student noted, "were the most entertaining talks. . . . He incited and gently forced others to think for themselves. . . ." Kant wrote and published the bulk of his work after he was named to the chair of logic and metaphysics at Konigsberg in 1770: *Critique of Pure Reason* (1781), *Critique of Practical Reason* (1788), and *Critique of Judgment* (1790). Kant was revered, but did not let fame alter his habits. He never traveled more than sixty miles from his home. Townspeople set their clocks by his precisely timed daily walk taken on a street named, in deference to him, "The Philosopher's Walk." Kant died in Konigsberg in 1804.

From *Lectures on Ethics,* translated by Louis Infield. Publisher: Methuen & Co., Ltd., 1930. Pages 129–135.

Conscience

Conscience is an instinct to pass judgment upon ourselves in accordance with moral laws. It is not a mere faculty, but an instinct; and its judgment is not logical, but judicial. We have the faculty to judge ourselves logically in terms of laws of morality; we can make such use as we please of this faculty. But conscience has the power to summon us against our will before the judgment-seat to be judged on account of the righteousness or unrighteousness of our actions. It is thus an instinct and not merely a faculty of judgment, and it is an instinct to judge, not in the logical, but in the judicial sense.

A judge passes judgment; he does not merely form a judgment. The difference is that he has the right to judge [with authority], and to give legal effect to his judgment. Thus his judgment has force of law, and is a sentence. The judge must either condemn or acquit, not merely form a judgment.

If our conscience were merely an impulse to form a judgment, it would be, like other faculties of which we are possessed (e.g., the impulses to compare ourselves with others, or to flatter ourselves), a faculty of knowledge. We all have an impulse to pat ourselves on the back for good actions done in accordance with rules of prudence. We likewise reproach ourselves for imprudent conduct. Thus we all have an impulse to flatter and blame ourselves in accordance with rules of prudence. But this tendency to praise and blame oneself is not conscience, though it is analogous to it and men often mistake it for conscience.

A criminal lying under sentence of death frets and worries and reproaches himself severely, but mainly for the imprudence which led to detection. He imagines that it is his conscience which reproaches him for his immorality; but it is not the pangs of conscience that he feels; for, had he got off scot-free, he would have felt no qualms, and if he had a conscience he would feel its reproach in any event. We must, therefore, differentiate between the judgment of prudence and the judgment of conscience.

Many people have only a semblance of conscience which they imagine is conscience itself. Death-bed repentance is a case in point. Such repentance is often enough not remorse for the immorality of behaviour, but for the folly of actions which, now that the judgment-seat is near, make it impossible to stand before the judge.

Vices bring their own punishments, and these punishments bring home to us the criminality of the acts; a man, therefore, who feels disgust with his past vices does not know whether his loathing is due to the punishments or to the criminality of his offences.

He who has no immediate loathing for what is morally wicked, and finds no pleasure in what is morally good, has no moral feeling, and such a man has no conscience. He who goes in fear of being prosecuted for a wicked deed, does not reproach himself on the score of the wickedness of his misdemeanour, but on the score of the painful consequences which await him; such a one has no conscience, but only a semblance of it. But he who has a sense of the wickedness of the deed itself, be the consequences what they may, has a conscience.

We must guard against confusion here: reproaches for the consequences of imprudence must not be confused with reproaches for breaches of morality. It is important in practice that a teacher, for instance, should look to see whether his pupil repents a deed from a true sense of its wickedness, or whether he feels remorse because he must soon face a judge before whom, on account of his action, he cannot hope to stand. Repentance which manifests itself for the first time on the death-bed has no moral worth: its motive is the nearness of death; if the approach of death were not feared there would probably be no repentance. The penitent in such a case may be likened to the unlucky gambler who fumes and rages against himself for his folly and tears his hair. He has no qualms about the vice; he hates its consequences. We ought not to be misled into consoling and comforting a man for such a semblance of conscience.

Prudence reproaches; conscience accuses. If a man has acted unwisely and reproaches himself for his imprudence no longer than is necessary for him to learn his lesson, he is observing a rule of prudence and it must be accounted to him for honour, for it is a sign of strength of character. But the accusation of conscience cannot be so readily dismissed, neither should it be; it is not a matter of the will, and the capacity to dismiss the accusation of a remorseful conscience is not evidence of strength of character, but rather of wickedness and religious impenitence. A man who can at will dismiss the accusations of conscience is a rebel, like the man who can disregard the accusation of his judge, over whom the judge has no power.

Conscience is an instinct to judge with legal authority according to moral laws; it pronounces a judicial verdict, and, like a judge who can only punish or acquit but cannot reward, so also our conscience either acquits or declares us guilty and deserving of punishment. The judgment has validity if it is felt and enforced. Two consequences follow from this. The first effectual expression of this judicial verdict which has the force of law is moral repentance; the second, without which the sentence is inoperative, is action in accordance with the judicial verdict. If it does not result in practical endeavour to do what is demanded for the satisfaction of the moral law, the conscience is but an idle conscience, and however penitent we may be the penitence is vain so long as we do not satisfy the debt we owe to the moral law; for even [in the human sphere] a debt is not satisfied by penitence, but by payment.

Preachers must, therefore, impress upon their hearers that, while they must repent for their transgressions against their duties to themselves, though they cannot remedy these, in the case of injustice done to others mere repentance is not enough: it must be followed by endeavour to remedy the injustice. Whining and lamentation are as useless [in the religious sphere] as they are [in the human sphere]. The history of death-bed repentances can, of course, show no instance of such practical repentance—a proof that it neglects an essential element.

The court of justice of our conscience can conveniently be compared with an ordinary law-court. We find in our hearts a prosecutor, for whom there would be no place unless there were also a law. This law, which is based on reason and not on sentiment, is incorruptible and incontestably just and pure; it is the moral law, established as the holy and inviolable law of humanity. Beside these there is equally an advocate within us, called Self-love, who brings forward many an argument in our defence, and whose pleas the prosecutor in his turn endeavours to refute. Lastly we find a judge within us who either condemns or acquits. It is impossible to blind his judgment. To refuse to appear before the bar of conscience is easier. Once we appear, the judge pronounces impartially, and his verdict falls normally upon the side of truth. If not, it must be because he judges by false principles of morality.

Except on the death-bed, when they listen more eagerly to the accuser, men lend a readier ear to their defender. A good conscience demands a pure law, for the accuser must be on the watch, whatever we do; and in judging our actions we must judge justly and morally and must have strength of conscience to give effect to the valid judgment. The conscience must have its operative, and not merely its speculative, principles; and to make its judgments operative it must be strong and command respect. Where is the judge who would be content to do no more than lecture and make judicial pronouncements? Judicial pronouncements must be put into operation.

Let us consider wherein the just conscience differs from one that is at fault. An error of conscience can be either an [error of fact or an error of law]. If a man's conscience errs and he acts in accordance with it, his acts may be at fault, but they cannot be accounted to him for a transgression. There are [blameworthy errors] and [innocent errors]. In respect of his natural obligations no man can be at fault; the natural moral laws must be known to all; they are contained in our reason; no man can, therefore, err innocently in respect of them, and in the case of natural laws there can be no innocent errors; but it is otherwise with positive laws; here we can have [innocent errors], and an [erring conscience] may give rise to actions which are not culpable. In respect of the natural law there can be no [innocent errors].

But what is a man to do when a positive and a natural law conflict? Take, for instance, the case of a man who is taught by his religion to execrate adherents of other religions, or that of a man who is told by Jesuits that good can come from knavery. Such a man would not be acting in accord with his conscience; the natural law is known to him, and he is aware that he ought on no account to act unrighteously. The verdict of natural conscience being in conflict with the verdict of instructed conscience, he must obey the former.

All positive laws are conditioned by the natural law, and they cannot, therefore, rightly contain anything which conflicts with it. To plead the excuse of an erring conscience is a serious matter, for in this way we could shift the responsibility for a great deal; but we are also accountable for our errors. . . .

Conscience is the representative within us of the divine judgment-seat: it weighs our dispositions and actions in the scales of a law which is holy and pure; we cannot deceive it, and, lastly, we cannot escape it because, like the divine omnipresence, it is always with us. Since, then, conscience is the representative within us of divine justice, we must not hurt or injure it.

We might contrast [natural conscience and artificial conscience]. Many have argued that conscience is a work of art and education, and that it judges and sentences by force of habit; but if this were the case, men with a conscience not so tutored and practised could escape the stings of conscience; there are, however, no cases of this.

It is obvious that art and instruction can only bring into fruition that for which we have a natural aptitude, so that if our conscience is to judge we must have a prior knowledge of good and evil. Yet a cultivated mind need not be followed by a cultivated conscience. Thus conscience is synonymous with natural conscience; and if we are to draw distinctions it should be between conscience before, during, and after the act.

Before the act the conscience has power enough to dissuade a man from committing the act, during the act it is stronger, and it is strongest of all after the act. Before the act conscience must still be weak; for the act has not yet been committed, and we feel weaker in the presence of an unsatisfied inclination which is still strong enough to withstand conscience. During the act conscience becomes stronger; and then, when inclination is satisfied, and has become too weak to withstand conscience, conscience is at its strongest. We can see from the case of passion that after satisfying his strongest desire, a man is even overcome by a feeling of disgust, because a strong passion, once satisfied, becomes too languid to offer opposition, so that conscience is at its strongest, and remorse follows. . . .

Conscience accompanying the act is gradually weakened by usage, so that in the long run man becomes as used to his vices as to tobacco smoke. In the end his conscience loses all its authority, the court of justice ceases to function and decide, and the accuser has no longer any task to perform.

Some burden their conscience with many matters of negligible importance . . .; they ask it to resolve problems of a quibbling nature, such as whether it is right to tell a lie in order to make an April fool of a person, or whether a rite or ceremony should be performed in this or that manner. This is casuistry, and it produces a micrological conscience. The subtler conscience is in such matters of detail, the worse is it in matters of practical importance, and people with such consciences are notoriously wont to concern themselves with speculations arising from positive law and to give themselves a free hand in everything else.

If a person is capable of reproaching himself for his sins, his conscience is said to be alive; but on the other hand, if a man searches needlessly for evidences of evil in his conduct, his conscience is melancholy. Conscience should not lord it over us like a tyrant; we do no hurt to our conscience by proceeding on our way cheerfully; tormenting consciences in the long run become dulled and ultimately cease to function.

KARL MARX was born in Trier, Germany in 1818 to
Jewish parents. His father was a lawyer who—probably
to gain professional acceptance—was baptized in the
Evangelical Established Church, as was Karl Marx.
Marx received his doctorate in philosophy from the
University of Jena in 1841. He then became editor of a
radical newspaper in Cologne. It was suppressed, and
soon afterwards Marx married and left for Paris. There
he met Friedrich Engels, who became Marx's lifelong
revolutionary collaborator. Marx's socialist views crystal-
lized, and he was expelled from Paris, Brussels (where
he and Engels wrote *The Communist Manifesto* in 1848),
Germany, and again from France for his political views
and activities. Marx helped to found the Communist
League, served as European correspondent for *The New
York Tribune,* and played a leading role in the International Working Men's Association. While impoverished
in London for a period of years, he studied history at
the British Museum and prepared *Capital* (1867, 1885,
1894). Marx died in London in 1883.

From *Karl Marx, Early Writings,* translated and edited
by T. B. Bottomore. Publisher: McGraw-Hill Book
Company, 1964. Pages 69–75 and 120–33.

Alienated Labour

Wages are determined by the bitter struggle between capitalist and worker. The necessary victory of the capitalist. The capitalist can live longer without the worker than can the worker without the capitalist. Combination among capitalists is usual and effective, whereas combination among workers is proscribed and has painful consequences for them. Moreover, the landowner and capitalist can supplement their revenues with the profits of industry, while the worker has neither ground rent nor interest on capital to add to his industrial earnings. Hence, the intensity of competition among workers. Only for the workers, therefore, is the separation of capital, landed property and labour an inescapable, vital and harmful separation. Capital and landed property need not remain in this abstraction, as must the labour of the workers.

For the worker, therefore, the separation of capital, ground rent and labour is fatal.

The lowest and the only necessary rate of wages is that which provides for the subsistence of the worker during work and for a supplement adequate to raise a family so that the race of workers does not die out. According to Smith, the normal wage is the lowest which is compatible with common humanity,[1] that is, with a bestial existence.

---

[1] [Adam Smith, *The Wealth of Nations.*]

The demand for men necessarily regulates the production of men, as of every other commodity. If the supply greatly exceeds the demand, then a section of the workers declines into beggary or starvation. Thus, the existence of the worker is reduced to the same conditions as the existence of any other commodity. The worker has become a commodity and he is fortunate if he can find a buyer. And the demand, upon which the worker's life depends, is determined by the caprice of the wealthy and the capitalists. If the supply exceeds the demand, one of the elements entering into price—profit, ground rent, wages—will be paid below its *rate;* a part of the supply of these factors will then be withdrawn from this use and the market price will gravitate towards the natural price. But (1) where there is an extensive division of labour it is extremely difficult for the worker to direct his labour into other uses, and (2) because of his subordination to the capitalist, he is the first to suffer hardship.

The worker, therefore, loses most, and loses inevitably, from the gravitation of the market price to the natural price. At the same time, it is the ability of the capitalist to put his capital to other uses which either condemns the worker, who is limited to one employment of his labour, to starvation, or forces him to accept every demand which the capitalist makes.

The adventitious and sudden variations in market price affect ground rent less than those parts of the price which comprise profit and wages, but they affect profit less than wages. In most cases, for every wage which rises there is one which remains *stationary* and one which *falls.*

The worker does not necessarily gain when the capitalist gains, but he necessarily loses with him. Thus, the worker does not gain if the capitalist succeeds in maintaining the market price above the natural price by means of manufacturing or commercial secret, a monopoly or the favourable situation of his property.

Further, *the prices of labour are much more stable than the prices of provisions.* They often vary inversely. In a dear year, wages fall, because of the decline in demand, but rise because of the increase in the price of provisions; so they balance. In any event, numbers of workers are without bread. In cheap years, wages rise because of increased demand, and fall because of the low prices of provisions; so they balance.

Another disadvantage of the worker: *The wage rates of different kinds of workers vary much more than do the profits in the different branches in which capital is employed.* In work, all the natural, spiritual, and social differences of individual activity appear and are differently remunerated, while dead capital maintains an unvarying performance and is indifferent to *real* individual activity.

In general, it should be noted that where worker and capitalist both suffer, the worker suffers in his existence while the capitalist suffers in the profit on his dead mammon.

The worker has not only to struggle for his physical means of subsistence; he must also struggle to obtain work, i.e., for the possibility and the means to perform his activity. Let us take three conditions in which society may find itself, and consider the situation of the worker in each of them.

1. If the wealth of society is diminishing, the worker suffers most, for although the working class cannot gain as much as the class of property owners in a prosperous state of society, *none suffers so cruelly from its decline as the working class.*

2. Let us next take a society in which wealth is increasing. This situation is the only one favourable to the worker. In this case, there is competition among capitalists and the demand for workers exceeds the supply. But, *in the first place,* the raising of wages leads to *overwork* among the workers. The more they want to earn the more they must sacrifice their time and perform slave labour in which their freedom is totally alienated in the service of avarice. In so doing they shorten their lives. This shortening of the life span is a favourable circumstance for the working class as a whole, since it makes necessary an ever renewed

supply of workers. This class must always sacrifice a part of itself, in order not to be ruined as a whole.

Furthermore, when is a society in a condition of increasing wealth? When the capital and revenues of a country are growing. But this is only possible (*a*) when much labour is accumulated, for capital is accumulated labour; when, therefore, more of the worker's product is taken from him, when his own labour becomes opposed to him as an alien possession, and when his means of existence and his activity are increasingly concentrated in the hands of the capitalist. (*b*) The accumulation of capital increases the division of labour, and the division of labour increases the number of workers; conversely, the increasing number of workers increases the division of labour, and the increasing division of labour increases the accumulation of capital. As a result of the division of labour on one hand, and the accumulation of capital on the other hand, the worker becomes even more completely dependent upon labour, and upon a particular, extremely one-sided, mechanical kind of labour. Just as he is reduced, therefore, both spiritually and physically to the condition of a machine, and from being a man becomes merely an abstract activity and a belly, so he becomes increasingly dependent upon all the fluctuations in market price, in the employment of capital, and in the caprices of the rich. Equally, the growth of the class of men who are entirely dependent upon work increases competition among the workers and lowers their price. In the factory system this situation of the workers reaches its climax.

(*c*) In a society where prosperity is increasing, only the very wealthiest can live from the interest on money. All others must employ their capital in business or trade. As a result the competition among capitalists increases. The concentration of capital becomes greater, the large capitalists ruin the small ones, and some of the former capitalists sink into the working class which, as a result of this accession of numbers, suffers a further decline in wages and falls into still greater dependence upon the few great capitalists. Since, at the same time, the number of capitalists has diminished, the competition among them for workers hardly exists any longer, whereas the competition among workers, on account of the increase in their numbers, has become greater, more abnormal and more violent. Consequently, a part of the working class falls into a condition of beggary or starvation, with the same necessity as a section of the middle capitalists falls into the working class.

Thus, even in the state of society which is most favourable to the worker, the inevitable result for the worker is overwork and premature death, reduction to a machine, enslavement to capital which accumulates in menacing opposition to him, renewed competition and beggary or starvation for a part of the workers.

Rising wages awake in the worker the same desire for enrichment as in the capitalist, but he can only satisfy it by the sacrifice of his body and spirit. Rising wages presuppose, and also bring about, the accumulation of capital; thus they increasingly alienate the product of labour from the worker. Likewise, the division of labour makes him increasingly one-sided and dependent, and introduces competition not only from other men but also from machines. Since the worker has been reduced to a machine, the machine can compete with him. Finally, just as the accumulation of capital increases the amount of industry, and thus the number of workers, so as a result of this accumulation the same volume of industry produces a *greater quantity of products* which leads to overproduction and culminates either in putting a great part of the workers out of work or in reducing their wages to the most wretched minimum. Such are the consequences of a state of society which is most favourable to the worker, namely a state of increasing, developing wealth.

Eventually, however, this state of growth must reach its culmination. What is then the condition of the worker?

3. "In a country which had acquired that full complement of riches . . . both the wages of labour and the profits of stock would probably be very low . . . the competition for employment would necessarily be so great as to reduce the wages of labour to what was barely sufficient to keep up the number of labourers, and, the country being already fully peopled, that number could never be augmented [Smith]." The excess would have to die.

Thus, in a declining state of society, increasing misery of the worker; in a progressive state, complicated misery; and in the final state, stationary misery.

Since, however, according to Smith a society is not happy in which the majority suffers, and since the wealthiest state of society leads to suffering for the majority, while the economic system (in general, a society of private interests) leads to this wealthiest state, it follows that social *misery* is the goal of the economy. . . .

ALIENATED LABOUR

We have begun from the presuppositions of political economy. We have accepted its terminology and its laws. We presupposed private property; the separation of labour, capital, and land, as also of wages, profit, and rent; the division of labour; competition; the concept of exchange value, etc. From political economy itself, in its own words, we have shown that the worker sinks to the level of a commodity, and to a most miserable commodity; that the misery of the worker increases with the power and volume of his production. . . .

Political economy begins with the fact of private property; it does not explain it. It conceives the *material* process of private property, as this occurs in reality, in general and abstract formulas which then serve it as laws. It does not *comprehend* these laws; that is, it does not show how they arise out of the nature of private property. Political economy provides no explanation of the basis for the distinction of labour from capital, of capital from land. When, for example, the relation of wages to profits is defined, this is explained in terms of the interests of capitalists; in other words, what should be explained is assumed. Similarly, competition is referred to at every point and is explained in terms of external conditions. Political economy tells us nothing about the extent to which these external and apparently accidental conditions are simply the expression of a necessary development. We have seen how exchange itself seems an accidental fact. The only motive forces which political economy recognizes are *avarice* and the *war between the avaricious, competition*. . . .

Let us not begin our explanation, as does the economist, from a legendary primordial condition. Such a primordial condition does not explain anything; it merely removes the question into a grey and nebulous distance. It asserts as a fact or event what it should deduce, namely, the necessary relation between two things; for example, between the division of labour and exchange. In the same way theology explains the origin of evil by the fall of man; that is, it asserts as a historical fact what it should explain.

We shall begin from a *contemporary* economic fact. The worker becomes poorer the more wealth he produces and the more his production increases in power and extent. The worker becomes an ever cheaper commodity the more goods he creates. The *devaluation* of the human world increases in direct relation with the *increase in value* of the world of things. Labour does not only create goods; it also produces itself and the worker as a *commodity,* and indeed in the same proportion as it produces goods.

This fact simply implies that the object produced by labour, its product, now stands opposed to it as an *alien being,* as a *power independent* of the producer. The product of labour is labour which has been embodied in an object and turned into a physical thing; this product is an *objectification* of labour. The performance of work is at the same time its objectification. The performance of work appears in the sphere of political economy as a *vitiation* of the worker, objectification as a *loss* and as *servitude to the object,* and appropriation as *alienation.*

So much does the performance of work appear as vitiation that the worker is vitiated to the point of starvation. So much does objectification appear as loss of the object that the worker is deprived of the most essential things not only of life but also of work. Labour itself becomes an object which he can acquire only by the greatest effort and with unpredictable interruptions. So much does the appropriation of the object appear as alienation that the more objects the worker produces the fewer he can possess and the more he falls under the domination of his product, of capital.

All these consequences follow from the fact that the worker is related to the *product of his labour* as to an *alien* object. For it is clear on this presupposition that the more the worker expends himself in work the more powerful becomes the world of objects which he creates in face of himself, the poorer he becomes in his inner life, and the less he belongs to himself. It is just the same as in religion. The more of himself man attributes to God the less he has left in himself. The worker puts his life into the object, and his life then belongs no longer to himself but to the object. The greater his activity, therefore, the less he possesses. What is embodied in the product of his labour is no longer his own. The greater this product is, therefore, the more he is diminished. The *alienation* of the worker in his product means not only that his labour becomes an object, assumes an *external* existence, but that it exists independently, *outside himself,* and alien to him, and that it stands opposed to him as an autonomous power. The life which he has given to the object sets itself against him as an alien and hostile force.

Let us now examine more closely the phenomenon of *objectification;* the worker's production and the *alienation* and *loss* of the object it produces, which is involved in it. The worker can create nothing without *nature,* without the *sensuous external world.* The latter is the material in which his labour is realized, in which it is active, out of which and through which it produces things.

But just as nature affords the *means of existence* of labour, in the sense that labour cannot *live* without objects upon which it can be exercised, so also it provides the *means of existence* in a narrower sense; namely the means of physical existence for the *worker* himself. Thus, the more the worker *appropriates* the external world of sensuous nature by his labour the more he deprives himself of *means of existence,* in two respects: first, that the sensuous external world becomes progressively less an object belonging to his labour or a means of existence of his labour, and secondly, that it becomes progressively less a means of existence in the direct sense, a means for the physical subsistence of the worker.

In both respects, therefore, the worker becomes a slave of the object; first, in that he receives an *object of work,* i.e., receives *work,* and secondly, in that he receives *means of subsistence.* Thus the object enables him to exist, first as a *worker* and secondly, as a *physical subject.* The culmination of this enslavement is that he can only maintain himself as a *physical subject* so far as he is a *worker,* and that it is only as a *physical subject* that he is a worker.

(The alienation of the worker in his object is expressed as follows in the laws of political economy: the more the worker produces the less he has to consume; the more value he creates the more worthless he becomes; the more refined his product the more crude and misshapen the worker; the more civilized the product the more barbarous the worker; the more powerful the work the more feeble the worker; the more the work manifests intelligence the more the worker declines in intelligence and becomes a slave of nature.)

Political economy conceals the alienation in the nature of labour insofar as it does not examine the direct relationship between the worker (work) and production. Labour certainly produces marvels for the rich but it produces privation for the worker. It produces palaces, but hovels for the worker. It produces beauty, but deformity for the worker. It replaces labour by machinery, but it casts some of the workers back into a barbarous kind of work and turns the others into machines. It produces intelligence, but also stupidity and cretinism for the workers.

The direct relationship of labour to its products is the relationship of the worker to the objects of his production. The relationship of property owners to the objects of production and to production itself is merely a *consequence* of this first relationship and confirms it. We shall consider this second aspect later.

Thus, when we ask what is the important relationship of labour, we are concerned with the relationship of the *worker* to production.

So far we have considered the alienation of the worker only from one aspect; namely, *his relationship with the products of his labour.* However, alienation appears not merely in the result but also in the *process of production,* within *productive activity* itself. How could the worker stand in an alien relationship to the product of his activity if he did not alienate himself in the act of production itself? The product is indeed only the *résumé* of activity, of production. Consequently, if the product of labour is alienation, production itself must be active alienation—the alienation of activity and the activity of alienation. The alienation of the object of labour merely summarizes the alienation in the work activity itself.

What constitutes the alienation of labour? First, that the work is *external* to the worker, that it is not part of his nature; and that, consequently, he does not fulfil himself in his work but denies himself, has a feeling of misery rather than well-being, does not develop freely his mental and physical energies but is physically exhausted and mentally debased. The worker, therefore, feels himself at home only during his leisure time, whereas at work he feels homeless. His work is not voluntary but imposed, *forced labour*. It is not the satisfaction of a need, but only a *means* for satisfying other needs. Its alien character is clearly shown by the fact that as soon as there is no physical or other compulsion it is avoided like the plague. External labour, labour in which man alienates himself, is a labour of self-sacrifice, of mortification. Finally, the external character of work for the worker is shown by the fact that it is not his own work but work for someone else, that in work he does not belong to himself but to another person.

Just as in religion the spontaneous activity of human fantasy, of the human brain and heart, reacts independently as an alien activity of gods or devils upon the individual, so the activity of the worker is not his own spontaneous activity. It is another's activity and a loss of his own spontaneity.

We arrive at the result that man (the worker) feels himself to be freely active only in his animal functions — eating, drinking, and procreating, or at most also in his dwelling and in personal adornment — while in his human functions he is reduced to an animal. The animal becomes human and the human becomes animal.

Eating, drinking, and procreating are of course also genuine human functions. But abstractly considered, apart from the environment of human activities, and turned into final and sole ends, they are animal functions.

We have now considered the act of alienation of practical human activity, labour, from two aspects: (1) the relationship of the worker to the *product of labour* as an alien object which dominates him. This relationship is at the same time the relationship to the sensuous external world, to natural objects, as an alien and hostile world; (2) the relationship of labour to the *act of production* within *labour*. This is the relationship of the worker to his own activity as something alien and not belonging to him, activity as suffering (passivity), strength as powerlessness, creation as emasculation, the *personal* physical and mental energy of the worker, his personal life (for what is life but activity?) as an activity which is directed against himself, independent of him and not belonging to him. This is *self-alienation* as against the above-mentioned alienation of the *thing*.

We have now to infer a third characteristic of *alienated labour* from the two we have considered.

Man is a species-being not only in the sense that he makes the community (his own as well as those of other things) his object both practically and theoretically, but also (and this is simply another expression for the same thing) in the sense that he treats himself as the present, living species, as a *universal* and consequently free being.

Species-life, for man as for animals, has its physical basis in the fact that man (like animals) lives from inorganic nature, and since man is more universal than an animal so the range of inorganic nature from which he lives is more universal. Plants, animals, minerals, air, light, etc., constitute, from the theoretical aspect, a part of human consciousness as objects of natural science and art; they are man's spiritual inorganic nature, his intellectual means of life, which he must first prepare for enjoyment and perpetuation. So also, from the practical aspect, they form a part of human life and activity. In practice man lives only from these natural products, whether in the form of food, heating, clothing, housing, etc. The universality of man appears in practice in the universality which makes the whole of nature into his inorganic body: (1) as a direct means of life; and equally (2) as the material object and instrument of his life activity. Nature is the inorganic body of man; that is to say nature, excluding the human body itself. To say that man *lives* from nature means that nature is his *body* with which he must remain in a continuous interchange in order not to die. The statement that the physical and mental life of man, and nature, are interdependent means simply that nature is interdependent with itself, for man is a part of nature.

Since alienated labour: (1) alienates nature from man; and (2) alienates man from himself, from his own active function, his life activity; so it alienates him from the species. It makes *species-life* into a means of individual life. In the first place it alienates species-life and individual life, and secondly, it turns the latter, as an abstraction, into the purpose of the former, also in its abstract and alienated form.

For labour, *life activity, productive life,* now appear to man only as *means* for the satisfaction of a need, the need to maintain his physical existence. Productive life is, however, species-life. It is life creating life. In the type of life activity resides the whole character of a species, its species-character; and free, conscious activity is the species-character of human beings. Life itself appears only as a *means of life.*

The animal is one with its life activity. It does not distinguish the activity from itself. It is *its activity.* But man makes his life activity itself an object of his will and consciousness. He has a conscious life activity. It is not a determination with which he is completely identified. Conscious life activity distinguishes man from the life activity of animals. Only for this reason is he a species-being. Or rather, he is only a self-conscious being, i.e., his own life is an object for him, because he is a species-being. Only for this reason is his activity free activity. Alienated labour reverses the relationship, in that man because he is a self-conscious being makes his life activity, his *being,* only a means for his *existence.*

The practical construction of an *objective world,* the *manipulation* of inorganic nature, is the confirmation of man as conscious species-being, i.e., a being who treats the species as his own being or himself as a species-being. Of course, animals also produce. They construct nests, dwellings, as in the case of bees, beavers, ants, etc. But they only produce what is strictly necessary for themselves or their young. They produce only in a single direction, while man produces universally. They produce only under the compulsion of direct physical needs, while man produces when he is free from physical need and only truly produces in freedom from such need. Animals produce only themselves, while man reproduces the whole of nature. The products of animal production belong directly to their physical bodies, while man is free in face of his product. Animals construct only in accordance with the standards and needs of the species to which they belong, while man knows how to produce in accordance with the standards of every species and knows how to apply the appropriate standard to the object. Thus man constructs also in accordance with the laws of beauty.

It is just in his work upon the objective world that man really proves himself as a *species-being*. This production is his active species-life. By means of it nature appears as *his* work and his reality. The object of labour is, therefore, the *objectification of man's species-life;* for he no longer reproduces himself merely intellectually, as in consciousness, but actively and in a real sense, and he sees his own reflection in a world which he has constructed. While, therefore, alienated labour takes away the object of production from man, it also takes away his *species-life,* his real objectivity as a species-being, and changes his advantage over animals into a disadvantage insofar as his inorganic body, nature, is taken from him.

Just as alienated labour transforms free and self-directed activity into a means, so it transforms the species-life of man into a means of physical existence.

Consciousness, which man has from his species, is transformed through alienation so that species-life becomes only a means for him. (3) Thus alienated labour turns the *species-life of man,* and also nature as his mental species-property, into an *alien* being and into a *means* for his *individual existence*. It alienates from man his own body, external nature, his mental life and his *human* life. (4) A direct consequence of the alienation of man from the product of his labour, from his life activity and from his species-life, is that *man* is *alienated* from other *men*. When man confronts himself he also confronts *other* men. What is true of man's relationship to his work, to the product of his work and to himself, is also true of his relationship to other men, to their labour and to the objects of their labour.

In general, the statement that man is alienated from his species-life means that each man is alienated from others, and that each of the others is likewise alienated from human life.

Human alienation, and above all the relation of man to himself, is first realized and expressed in the relationship between each man and other men. Thus in the relationship of alienated labour every man regards other men according to the standards and relationships in which he finds himself placed as a worker.

We began with an economic fact, the alienation of the worker and his production. We have expressed this fact in conceptual terms as *alienated labour,* and in analysing the concept we have merely analysed an economic fact.

Let us now examine further how this concept of alienated labour must express and reveal itself in reality. If the product of labour is alien to me and confronts me as an alien power, to whom does it belong? If my own activity does not belong to me but is an alien, forced activity, to whom does it belong? To a being *other* than myself. And who is this being? The *gods?* It is apparent in the earliest stages of advanced production, e.g., temple building, etc., in Egypt, India, Mexico, and in the service rendered to gods, that the product belonged to the gods. But the gods alone were never the lords of labour. And no more was *nature*. What a contradiction it would be if the more man subjugates nature by his labour, and the more the marvels of the gods are rendered superfluous by the marvels of industry, the more he should abstain from his joy in producing and his enjoyment of the product for love of these powers.

The *alien* being to whom labour and the product of labour belong, to whose service labour is devoted, and to whose enjoyment the product of labour goes, can only be *man* himself. If the product of labour does not belong to the worker, but confronts him as an alien power, this can only be because it belongs to *a man other than the worker*. If his activity is a torment to him it must be a source of *enjoyment* and pleasure to another. Not the gods, nor nature, but only man himself can be this alien power over men.

Consider the earlier statement that the relation of man to himself is first *realized, objectified,* through his relation to other men. If he is related to the product of his labour, his objectified labour, as to an *alien,* hostile, powerful and independent object, he is related in such a way that another alien, hostile, powerful and independent man is the lord of this object. If he is related to his own activity as to unfree activity, then he is related to it as activity in the service, and under the domination, coercion and yoke, of another man.

Every self-alienation of man, from himself and from nature, appears in the relation which he postulates between other men and himself and nature. Thus religious self-alienation is necessarily exemplified in the relation between laity and priest, or, since it is here a question of the spiritual world, between the laity and a mediator. In the real world of practice this self-alienation can only be expressed in the real, practical relation of man to his fellow men. The medium through which alienation occurs is itself a *practical* one. Through alienated labour, therefore, man not only produces his relation to the object and to the process of production as to alien and hostile men; he also produces the relation of other men to his production and his product, and the relation between himself and other men. Just as he creates his own production as a vitiation, a punishment, and his own product as a loss, as a product which does not belong to him, so he creates the domination of the non-producer over production and its product. As he alienates his own activity, so he bestows upon the stranger an activity which is not his own.

Thus, through alienated labour the worker creates the relation of another man, who does not work and is outside the work process, to this labour. The relation of the worker to work also produces the relation of the capitalist (or whatever one likes to call the lord of labour) to work. *Private property* is, therefore, the product, the necessary result, of *alienated labour,* of the external relation of the worker to nature and to himself.

Private property is thus derived from the analysis of the concept of *alienated labour;* that is, alienated man, alienated labour, alienated life, and estranged man.

We have, of course, derived the concept of *alienated labour (alienated life)* from political economy, from an analysis of the *movement of private property.* But the analysis of this concept shows that although private property appears to be the basis and cause of alienated labour, it is rather a consequence of the latter, just as the gods are *fundamentally* not the cause but the product of confusions of human reason. At a later stage, however, there is a reciprocal influence.

Only in the final stage of the development of private property is its secret revealed, namely, that it is on one hand the *product* of alienated labour, and on the other hand the *means* by which labour is alienated, *the realization of this alienation.* . . .

As we have discovered the concept of *private property* by an *analysis* of the concept of *alienated labour,* so with the aid of these two factors we can evolve all the *categories* of political economy, and in every category, e.g., trade, competition, capital, money, we shall discover only a particular and developed expression of these fundamental elements.

From "The First Book of Moses, called Genesis" in *The Holy Bible,* King James Version. Publisher: Meridian Books, The World Publishing Co., 1964. Chapters: 1; 2; 3; 4; 5:28–32; 6; 7; 8; 9; 11:1–9; 12; 14:1,2, 11–24; 15; 16; 17; 18; 19; 20; 21; 22:1–19; 23.

Genesis

THE CREATION, THE FLOOD, ABRAHAM

In the beginning God created the heaven and the earth. And the earth was without form, and void; and darkness was upon the face of the deep. And the Spirit of God moved upon the face of the waters. And God said, Let there be light: and there was light. And God saw the light, that it was good: and God divided the light from the darkness. And God called the light Day, and the darkness he called Night. And the evening and the morning were the first day.

And God said, Let there be a firmament in the midst of the waters, and let it divide the waters from the waters. And God made the firmament and divided the waters which were under the firmament from the waters which were above the firmament: and it was so. And God called the firmament Heaven. And the evening and the morning were the second day.

And God said, Let the waters under the heaven be gathered together unto one place, and let the dry land appear: and it was so. And God called the dry land Earth; and the gathering together of the waters called he Seas: and God saw that it was good. And God said, Let the earth bring forth grass, the herb yielding seed, and the fruit tree yielding fruit after his kind, whose seed is in itself, upon the earth: and it was so. And the earth brought forth grass, and herb yielding seed after his kind, and the tree yielding fruit, whose seed was in itself, after his kind: and God saw that it was good. And the evening and the morning were the third day.

And God said, Let there be lights in the firmament of the heaven to divide the day from the night; and let them be for signs, and for seasons, and for days, and years; and let them be for lights in the firmament of the heaven to give light upon the earth: and it was so. And God made two great lights; the greater light to rule the day, and the lesser light to rule the night: he made the stars also. And God set them in the firmament of the heaven to give light upon the earth, and to rule over the day and over the night, and to divide the light from the darkness: and God saw that it was good. And the evening and the morning were the fourth day. And God said, Let the waters bring forth abundantly the moving creature that hath life, and fowl that may fly above the earth in the open firmament of heaven. And God created great whales, and every living creature that moveth, which the waters brought forth abundantly, after their kind, and every winged fowl after his kind: and God saw that it was good. And God blessed them, saying, Be fruitful, and multiply, and fill the waters in the seas, and let fowl multiply in the earth. And the evening and the morning were the fifth day.

And God said, Let the earth bring forth the living creature after his kind, cattle, and creeping thing, and beast of the earth after his kind: and it was so. And God made the beast of the earth after his kind, and cattle after their kind, and every thing that creepeth upon the earth after his kind: and God saw that it was good.

And God said, Let us make man in our image, after our likeness: and let them have dominion over the fish of the sea, and over the fowl of the air, and over the cattle, and over all the earth, and over every creeping thing that creepeth upon the earth. So God created man in his own image, in the image of God created he him; male and female created he them. And God blessed them, and God said unto them, Be fruitful, and multiply, and replenish the earth, and subdue it: and have dominion over the fish of the sea, and over the fowl of the air, and over every living thing that moveth upon the earth.

And God said, Behold, I have given you every herb bearing seed, which is upon the face of all the earth, and every tree, in which is the fruit of a tree yielding seed; to you it shall be for meat. And to every beast of the earth, and to every fowl of the air, and to every thing that creepeth upon the earth, wherein there is life, I have given every green herb for meat: and it was so. And God saw every thing that he had made, and, behold, it was very good. And the evening and the morning were the sixth day.

Thus the heavens and the earth were finished, and all the host of them. And on the seventh day God ended his work which he had made; and he rested on the seventh day from all his work which he had made. And God blessed the seventh day, and sanctified it: because that in it he had rested from all his work which God created and made.

These are the generations of the heavens and of the earth when they were created, in the day that the Lord God made the earth and the heavens, and every plant of the field before it was in the earth, and every herb of the field before it grew: for the Lord God had not caused it to rain upon the earth, and there was not a man to till the ground. But there went up a mist from the earth, and watered the whole face of the ground. And the Lord God formed man of the dust of the ground, and breathed into his nostrils the breath of life; and man became a living soul.

And the Lord God planted a garden eastward in Eden; and there he put the man whom he had formed. And out of the ground made the Lord God to grow every tree that is pleasant to the sight, and good for food; the tree of life also in the midst of the garden, and the tree of knowledge of good and evil. And a river went out of Eden to water the garden; and from thence it was parted, and became into four heads. The name of the first is Pison: that is it which compasseth the whole land of Havilah, where there is gold; and the gold of that land is good: there is bdellium and the onyx stone. And the name of the second river is Gihon: the same is it that compasseth the whole land of Ethiopia. And the name of the third river is Hiddekel: that is it which goeth toward the east of Assyria. And the fourth river is Euphrates. And the Lord God took the man, and put him into the garden of Eden to dress it and to keep it. And the Lord God commanded the man, saying, Of every tree of the garden thou mayest freely eat; but of the tree of the knowledge of good and evil, thou shalt not eat of it: for in the day that thou eatest thereof thou shalt surely die.

And the Lord God said, It is not good that the man should be alone; I will make him an help meet for him. And out of the ground the Lord God formed every beast of the field, and every fowl of the air; and brought them unto Adam to see what he would call them: and whatsoever Adam called every living creature, that was the name thereof. And Adam gave names to all cattle, and to the fowl of the air, and to every beast of the field; but for Adam there was not found an help meet for him. And the Lord God caused a deep sleep to fall upon Adam, and he slept: and he took one of his ribs, and closed up the flesh instead thereof; and the rib, which the Lord God had taken from man, made he a woman, and brought her unto the man. And Adam said, This is now bone of my bones, and flesh of my flesh: she shall be called Woman, because she was taken out of Man. Therefore shall a man leave his father and his mother, and shall cleave unto his wife: and they shall be one flesh. And they were both naked, the man and his wife, and were not ashamed.

Now the serpent was more subtle than any beast of the field which the Lord God had made. And he said unto the woman, Yea, hath God said, Ye shall not eat of every tree of the garden? And the woman said unto the serpent, We may eat of the fruit of the trees of the garden; but of the fruit of the tree which is in the midst of the garden, God hath said, Ye shall not eat of it, neither shall ye touch it, lest ye die. And the serpent said unto the woman, Ye shall not surely die. For God doth know that in the day ye eat thereof, then your eyes shall be opened, and ye shall be as gods, knowing good and evil. And when the woman saw that the tree was good for food, and that it was pleasant to the eyes, and a tree to be desired to make one wise, she took of the fruit thereof, and did eat, and gave also unto her husband with her; and he did eat. And the eyes of them both were opened, and they knew that they were naked; and they sewed fig leaves together, and made themselves aprons.

And they heard the voice of the Lord God walking in the garden in the cool of the day: and Adam and his wife hid themselves from the presence of the Lord God amongst the trees of the garden. And the Lord God called unto Adam, and said unto him, Where art thou? And he said, I heard thy voice in the garden, and I was afraid, because I was naked; and I hid myself. And he said, Who told thee that thou wast naked? Hast thou eaten of the tree, whereof I commanded thee that thou shouldest not eat? And the man said, The woman whom thou gavest to be with me, she gave me of the tree, and I did eat. And the Lord God said unto the woman, What is this that thou hast done? And the woman said, The serpent beguiled me, and I did eat. And the Lord God said unto the serpent, Because thou hast done this, thou art cursed above all cattle, and above every beast of the field; upon thy belly shalt thou go, and dust shalt thou eat all the days of thy life. And I will put enmity between thee and the woman, and between thy seed and her seed; it shall bruise thy head, and thou shalt bruise his heel.

Unto the woman he said, I will greatly multiply thy sorrow and thy conception; in sorrow thou shalt bring forth children; and thy desire shall be to thy husband, and he shall rule over thee. And unto Adam he said, Because thou has hearkened unto the voice of thy wife, and hast eaten of the tree, of which I commanded thee, saying, Thou shalt not eat of it: cursed is the ground for thy sake; in sorrow shalt thou eat of it all the days of thy life; thorns also and thistles shall it bring forth to thee; and thou shalt eat the herb of the field; in the sweat of thy face shalt thou eat bread, till thou return unto the ground; for out of it wast thou taken: for dust thou art, and unto dust shalt thou return.

And Adam called his wife's name Eve; because she was the mother of all living. Unto Adam also and to his wife did the Lord God make coats of skins, and clothed them.

And the Lord God said, Behold, the man is become as one of us, to know good and evil: and now, lest he put forth his hand, and take also of the tree of life, and eat, and live for ever: therefore the Lord God sent him forth from the garden of Eden, to till the ground from whence he was taken. So he drove out the man; and he placed at the east of the garden of Eden Cherubims, and a flaming sword which turned every way, to keep the way of the tree of life.

And Adam knew Eve his wife; and she conceived, and bare Cain, and said, I have gotten a man from the Lord. And she again bare his brother Abel. And Abel was a keeper of sheep, but Cain was a tiller of the ground. And in process of time it came to pass, that Cain brought of the fruit of the ground an offering unto the Lord. And Abel, he also brought of the firstlings of his flock and of the fat thereof. And the Lord had respect unto Abel and to his offering: but unto Cain and to his offering he had not respect. And Cain was very wroth, and his countenance fell. And the Lord said unto Cain, Why art thou wroth?

and why is thy countenance fallen? If thou doest well, shalt thou not be accepted? and if thou doest not well, sin lieth at the door. And unto thee shall be his desire, and thou shalt rule over him. And Cain talked with Abel his brother: and it came to pass, when they were in the field, that Cain rose up against Abel his brother, and slew him.

And the Lord said unto Cain, Where is Abel thy brother? And he said, I know not: Am I my brother's keeper? And he said, What hast thou done? the voice of thy brother's blood crieth unto me from the ground. And now art thou cursed from the earth, which hath opened her mouth to receive thy brother's blood from thy hand. When thou tillest the ground, it shall not henceforth yield unto thee her strength; a fugitive and a vagabond shalt thou be in the earth. And Cain said unto the Lord, My punishment is greater than I can bear. Behold, thou hast driven me out this day from the face of the earth; and from thy face shall I be hid; and I shall be a fugitive and a vagabond in the earth; and it shall come to pass, that every one that findeth me shall slay me. And the Lord said unto him, Therefore whosoever slayeth Cain, vengeance shall be taken on him sevenfold. And the Lord set a mark upon Cain, lest any finding him should kill him.

And Cain went out from the presence of the Lord, and dwelt in the land of Nod, on the east of Eden. And Cain knew his wife; and she conceived, and bare Enoch: and he builded a city and called the name of the city after the name of his son, Enoch. And unto Enoch was born Irad: and Irad begat Mehujael: and Mehujael begat Methusael: and Methusael begat Lamech.

And Lamech took unto him two wives: the name of the one was Adah, and the name of the other Zillah. And Adah bare Jabal: he was the father of such as dwell in tents, and of such as have cattle. And his brother's name was Jubal: he was the father of all such as handle the harp and organ. And Zillah, she also bare Tubalcain, an instructor of every artificer in brass and iron: and the sister of Tubalcain was Naamah. And Lamech said unto his wives, Adah and Zillah, Hear my voice; ye wives of Lamech, hearken unto my speech: for I have slain a man to my wounding, and a young man to my hurt. If Cain shall be avenged sevenfold, truly Lamech seventy and sevenfold.

And Adam knew his wife again; and she bare a son, and called his name Seth: For God, said she, hath appointed me another seed instead of Abel, whom Cain slew. And to Seth, to him also there was born a son; and he called his name Enos: then began men to call upon the name of the Lord. . . .

[Here the descendents of Seth are traced through seven generations to Lamech.]

And Lamech lived an hundred eighty and two years, and begat a son: and he called his name Noah, saying, This same shall comfort us concerning our work and toil of our hands, because of the ground which the Lord hath cursed. And Lamech lived after he begat Noah five hundred ninety and five years, and begat sons and daughters. And all the days of Lamech were seven hundred seventy and seven years: and he died. And Noah was five hundred years old: and Noah begat Shem, Ham, and Japheth.

And it came to pass, when men began to multiply on the face of the earth, and daughters were born unto them, that the sons of God saw the daughters of men that they were fair; and they took them wives of all which they chose. And the Lord said, My spirit shall not always strive with man, for that he also is flesh: yet his days shall be an hundred and twenty years. There were giants in the earth in those days; and also after that, when the sons of God came in unto the daughters of men, and they bare children to them; the same became mighty men which were of old, men of renown.

And God saw that the wickedness of man was great in the earth, and that every imagination of the thoughts of his heart was only evil continually. And it repented the Lord that he had made man on the earth, and it grieved him at his heart. And the Lord said, I will destroy man whom I have created from the face of the earth; both man, and beast, and the creeping thing, and the fowls of the air; for it repenteth me that I have made them. But Noah found grace in the eyes of the Lord.

These are the generations of Noah: Noah was a just man and perfect in his generations, and Noah walked with God. And Noah begat three sons, Shem, Ham, and Japheth. The earth also was corrupt before God, and the earth was filled with violence. And God looked upon the earth, and, behold, it was corrupt; for all flesh had corrupted his way upon the earth. And God said unto Noah, The end of all flesh is come before me; for the earth is filled with violence through them; and, behold, I will destroy them with the earth.

Make thee an ark of gopher wood; rooms shalt thou make in the ark, and shalt pitch it within and without with pitch. And this is the fashion which thou shalt make it of: The length of the ark shall be three hundred cubits, the breadth of it fifty cubits, and the height of it thirty cubits. A window shalt thou make to the ark, and in a cubit shalt thou finish it above; and the door of the ark shalt thou set in the side thereof; with lower, second, and third stories shalt thou make it. And, behold, I, even I, do bring a flood of waters upon the earth, to destroy all flesh, wherein is the breath of life, from under heaven; and every thing that is in the earth shall die. But with thee will I establish my covenant; and thou shalt come into the ark, thou, and thy sons, and thy wife, and thy sons' wives with thee. And of every living thing of all flesh, two of every sort shalt thou bring into the ark, to keep them alive with thee; they shall be male and female. Of fowls after their kind, and of cattle after their kind, of every creeping thing of the earth after his kind, two of every sort shall come unto thee, to keep them alive. And take thou unto thee of all food that is eaten, and thou shalt gather it to thee; and it shall be for food for thee, and for them. Thus did Noah according to all that God commanded him, so did he.

And the Lord said unto Noah, Come thou and all thy house into the ark; for thee have I seen righteous before me in this generation. Of every clean beast thou shalt take to thee by sevens, the male and his female: and of beasts that are not clean by two, the male and his female. Of fowls also of the air by sevens, the male and the female; to keep seed alive upon the face of all the earth. For yet seven days, and I will cause it to rain upon the earth forty days and forty nights; and every living substance that I have made will I destroy from off the face of the earth. And Noah did according unto all that the Lord commanded him. And Noah was six hundred years old when the flood of waters was upon the earth.

And Noah went in, and his sons, and his wife, and his sons' wives with him, into the ark, because of the waters of the flood. Of clean beasts, and of beasts that are not clean, and of fowls, and of every thing that creepeth unto the earth, there went in two and two unto Noah into the ark, the male and the female, as God had commanded Noah. And it came to pass after seven days, that the waters of the flood were upon the earth.

In the six hundredth year of Noah's life, in the second month, the seventeenth day of the month, the same day were all the fountains of the great deep broken up, and the windows of heaven were opened. And the rain was upon the earth forty days and forty nights. In the selfsame day entered Noah, and Shem, and Ham, and Japheth, and sons of Noah, and Noah's wife, and the three wives of his sons with them, into the ark; they, and every beast after his kind, and all the cattle after their kind, and every creeping thing that creepeth upon the earth after his kind, and every fowl after his kind, every bird of every sort. And they went in unto Noah into the ark, two and two of all flesh, wherein is the breath of life. And they that went in, went in male and female of all flesh, as God had commanded him: and the Lord shut him in.

And the flood was forty days upon the earth; and the waters increased, and bare up the ark, and it was lift up above the earth. And the waters prevailed, and were increased greatly upon the earth; and the ark went upon the face of the waters. And the waters prevailed exceedingly upon the earth; and all the high hills, that were under the whole heaven, were covered. Fifteen cubits upward did the waters prevail; and the mountains were covered. And all the flesh died that moved upon the earth, both of fowl, and of cattle, and of beast, and of every creeping thing that creepeth upon the earth, and every man: All in whose nostrils was the breath of life, of all that was in the dry land, died. And every living substance was destroyed which was upon the face of the ground, both man, and cattle, and the creeping things, and the fowl of the heaven; and they were destroyed from the earth: and Noah only remained alive, and they that were with him in the ark. And the waters prevailed upon the earth an hundred and fifty days.

And God remembered Noah, and every living thing, and all the cattle that was with him in the ark: and God made a wind to pass over the earth, and the waters assuaged; the fountains also of the deep and the windows of heaven were stopped, and the rain from heaven was restrained; and the waters returned from off the earth continually: and after the end of the hundred and fifty days the waters were abated. And the ark rested in the seventh month, on the seventeenth day of the month, upon the mountains of Ararat. And the waters decreased continually until the tenth month: in the tenth month, on the first day of the month, were the tops of the mountains seen.

And it came to pass at the end of forty days, that Noah

opened the window of the ark which he had made: and he sent
forth a raven, which went forth to and fro, until the waters were
dried up from off the earth. Also he sent forth a dove from
him, to see if the waters were abated from off the face of the
ground; but the dove found no rest for the sole of her foot, and
she returned unto him into the ark, for the waters were on the
face of the whole earth: then he put forth his hand, and took
her, and pulled her in unto him into the ark. And he stayed
yet other seven days; and again he sent forth the dove out of
the ark; and the dove came in to him in the evening; and, lo,
in her mouth was an olive leaf pluckt off: so Noah knew that
the waters were abated from off the earth. And he stayed yet
other seven days; and sent forth the dove; which returned not
again unto him any more.

And it came to pass in the six hundredth and first year, in
the first month, the first day of the month, the waters were
dried up from off the earth: and Noah removed the covering
of the ark, and looked, and, behold, the face of the ground was
dry. And in the second month, on the seven and twentieth day
of the month, was the earth dried.

And God spake unto Noah, saying, Go forth of the ark,
thou, and thy wife, and thy sons, and thy sons' wives with thee.
Bring forth with thee every living thing that is with thee, of all
flesh, both of fowl, and of cattle, and of every creeping thing
that creepeth upon the earth; that they may breed abundantly
in the earth, and be fruitful, and multiply upon the earth. And
Noah went forth, and his sons, and his wife, and his sons' wives
with him: every beast, every creeping thing, and every fowl, and
whatsoever creepeth upon the earth, after their kinds, went forth
out of the ark.

And Noah builded an altar unto the Lord; and took of every
clean beast, and of every clean fowl, and offered burnt offerings
on the altar. And the Lord smelled a sweet savour; and the Lord
said in his heart, I will not again curse the ground any more

for man's sake; for the imagination of man's heart is evil from his youth; neither will I again smite any more every thing living, as I have done. While the earth remaineth, seedtime and harvest, and cold and heat, and summer and winter, and day and night shall not cease.

And God blessed Noah and his sons, and said unto them, Be fruitful, and multiply, and replenish the earth. And the fear of you and the dread of you shall be upon every beast of the earth, and upon every fowl of the air, upon all that moveth upon the earth, and upon all the fishes of the sea; into your hand are they delivered. Every moving thing that liveth shall be meat for you; even as the green herb have I given you all things. But flesh with the life thereof, which is the blood thereof, shall ye not eat. And surely your blood of your lives will I require; at the hand of every beast will I require it, and at the hand of man; at the hand of every man's brother will I require the life of man. Whoso sheddeth man's blood, by man shall his blood be shed: for in the image of God made he man. And you, be ye fruitful, and multiply; bring forth abundantly in the earth, and multiply therein.

And God spake unto Noah, and to his sons with him, saying, And I, behold, I establish my covenant with you, and with your seed after you; and with every living creature that is with you, of the fowl, of the cattle, and of every beast of the earth with you; from all that go out of the ark, to every beast of the earth. And I will establish my covenant with you; neither shall all flesh be cut off any more by the waters of a flood; neither shall there any more be a flood to destroy the earth. And God said, This is the token of the covenant which I make between me and you and every living creature that is with you, for perpetual generations: I do set my bow in the cloud, and it shall be for a token of a covenant between me and the earth. And it shall come to pass, when I bring a cloud over the earth, that the bow shall be seen in the cloud: and I will remember my covenant, which is between me and you and every living creature of all flesh; and the waters shall no more become a flood to destroy all flesh. And the bow shall be in the cloud; and I will look upon it, that I may remember the everlasting covenant between God and every living creature of all flesh that is upon the earth. And God said unto Noah, This is the token of the covenant, which I have established between me and all flesh that is upon the earth.

And the sons of Noah, that went forth of the ark, were Shem, and Ham, and Japheth: and Ham is the father of Canaan. These are the three sons of Noah: and of them was the whole earth overspread. And Noah began to be an husbandman, and he planted a vineyard: And he drank of the wine, and was drunken; and he was uncovered within his tent. And Ham, the father of Canaan, saw the nakedness of his father, and told his two brethren without. And Shem and Japheth took a garment, and laid it upon both their shoulders, and went backward, and covered the nakedness of their father; and their faces were backward, and they saw not their father's nakedness. And Noah awoke from his wine, and knew what his younger son had done unto him. And he said, Cursed be Canaan; a servant of servants shall he be unto his brethren. And he said, Blessed be the Lord God of Shem; and Canaan shall be his servant. God shall enlarge Japheth, and he shall dwell in the tents of Shem; and Canaan shall be his servant.

And Noah lived after the flood three hundred and fifty years. And all the days of Noah were nine hundred and fifty years: and he died. . . .

[Here the families of the sons of Noah are listed, and the nations of the world after the flood are traced to them.]

And the whole earth was of one language, and of one speech.

And it came to pass, as they journeyed from the east, that they found a plain in the land of Shinar; and they dwelt there. And they said one to another, Go to, let us make brick, and burn them throughly. And they had brick for stone, and slime had they for mortar. And they said, Go to, let us build us a city and a tower, whose top may reach unto heaven; and let us make us a name, lest we be scattered abroad upon the face of the whole earth. And the Lord came down to see the city and the tower, which the children of men builded. And the Lord said, Behold, the people is one, and they have all one language; and this they begin to do: and now nothing will be restrained from them, which they have imagined to do. Go to, let us go down, and there confound their language, that they may not understand one another's speech. So the Lord scattered them abroad from thence upon the face of all the earth: and they left off to build the city. Therefore is the name of it called Babel; because the Lord did there confound the language of all the earth: and from thence did the Lord scatter them abroad upon the face of all the earth. . . .

[Here the descendants of Shem are traced through seven generations to Abram and his nephew Lot.]

Now the Lord had said unto Abram, Get thee out of thy country, and from thy kindred, and from thy father's house, unto a land that I will shew thee: and I will make of thee a great nation, and I will bless thee, and make thy name great; and thou shalt be a blessing: and I will bless them that bless thee, and curse him that curseth thee: and in thee shall all families of the earth be blessed. So Abram departed, as the Lord had spoken unto him: and Lot went with him: and Abram was seventy and five years old when he departed out of Haran. And Abram took Sarai his wife, and Lot his brother's son, and all their substance that they had gathered, and the souls that they had gotten in

Haran; and they went forth to go into the land of Canaan; and into the land of Canaan they came.

And Abram passed through the land unto the place of Sichem, unto the plain of Moreh. And the Canaanite was then in the land. And the Lord appeared unto Abram; and said, Unto thy seed will I give this land: and there builded he an altar unto the Lord, who appeared unto him. And he removed from thence unto a mountain on the east of Beth-el, and pitched his tent, having Beth-el on the west, and Hai on the east: and there he builded an altar unto the Lord, and called upon the name of the Lord. And Abram journeyed, going on still toward the south.

And there was a famine in the land: and Abram went down into Egypt to sojourn there; for the famine was grievous in the land. And it came to pass, when he was come near to enter into Egypt, that he said unto Sarai his wife, Behold now, I know that thou art a fair woman to look upon: therefore it shall come to pass, when the Egyptians shall see thee, that they shall say, This is his wife: and they will kill me, but they will save thee alive. Say, I pray thee, thou art my sister: that it may be well with me for thy sake; and my soul shall live because of thee.

And it came to pass, that, when Abram was come into Egypt, the Egyptians beheld the woman that she was very fair. The princes also of Pharaoh saw her, and commended her before Pharaoh: and the woman was taken into Pharaoh's house. And he entreated Abram well for her sake: and he had sheep, and oxen, and he-asses, and menservants, and maidservants, and she-asses, and camels. And the Lord plagued Pharaoh and his house with great plagues because of Sarai Abram's wife. And Pharaoh called Abram, and said, What is this that thou hast done unto me? why didst thou not tell me that she was thy wife? Why saidst thou, She is my sister? so I might have taken her to me to wife: now therefore behold thy wife, take her, and go thy way. And Pharaoh commanded his men concerning him: and they sent him away, and his wife, and all that he had.

And Abram went up out of Egypt, he, and his wife, and all that he had, and Lot with him, into the south. And Abram was very rich in cattle, in silver, and in gold. And he went on his journeys from the south even to Beth-el, unto the place where his tent had been at the beginning, between Beth-el and Hai; unto the place of the altar, which he had made there at the first: and there Abram called on the name of the Lord.

And Lot also, which went with Abram, had flocks, and herds, and tents. And the land was not able to bear them, that they might dwell together: for their substance was great, so that they could not dwell together. And there was a strife between the herdmen of Abram's cattle and the herdmen of Lot's cattle: and the Canaanite and the Perizzite dwelled then in the land. And Abram said unto Lot, Let there be no strife, I pray thee, between me and thee, and between my herdmen and thy herdmen; for we be brethren. Is not the whole land before thee? separate thyself, I pray thee, from me: if thou wilt take the left hand, then I will go to the right; or if thou depart to the right hand, then I will go to the left. And Lot lifted up his eyes, and beheld all the plain of Jordan, that it was well watered every where, before the Lord destroyed Sodom and Gomorrah, even as the garden of the Lord, like the land of Egypt, as thou comest unto Zoar. Then Lot chose him all the plain of Jordan; and Lot journeyed east: and they separated themselves the one from the other. Abram dwelled in the land of Canaan, and Lot dwelled in the cities of the plain, and pitched his tent toward Sodom. But the men of Sodom were wicked and sinners before the Lord exceedingly.

And the Lord said unto Abram, after that Lot was separated from him, Lift up now thine eyes, and look from the place where thou art northward, and southward, and eastward, and westward: for all the land which thou seest, to thee will I give it, and to thy seed for ever. And I will make thy seed as the dust of the earth: so that if a man can number the dust of the earth, then shall thy seed also be numbered. Arise, walk through the land in the length of it and in the breadth of it; for I will give it unto thee. Then Abram removed his tent, and came and dwelt in the plain of Mamre, which is in Hebron, and built there an altar unto the Lord.

And it came to pass in the days of Amraphel king of Shinar, Arioch king of Ellasar, Chedorlaomer king of Elam, and Tidal king of nations; that these made war with Bera king of Sodom, and with Birsha king of Gomorrah, Shinab king of Admah, and Shemeber king of Zeboiim, and the king of Bela, which is Zoar. . . . And they took all the goods of Sodom and Gomorrah, and all their victuals, and went their way. And they took Lot, Abram's brother's son, who dwelt in Sodom, and his goods, and departed.

And there came one that had escaped, and told Abram the Hebrew; for he dwelt in the plain of Mamre the Amorite, brother of Eshcol, and brother of Aner: and these were confederate with Abram. And when Abram heard that his brother was taken captive, he armed his trained servants, born in his own house, three hundred and eighteen, and pursued them unto Dan. And he divided himself against them, he and his servants, by night, and smote them, and pursued them unto Hobah, which is on the left hand of Damascus. And he brought back all the goods, and also brought again his brother Lot, and his goods, and the women also, and the people.

And the king of Sodom went out to meet him after his return from the slaughter of Chedorlaomer, and of the kings that were with him, at the valley of Shaveh, which is the king's dale. And Melchizedek king of Salem brought forth bread and wine: and he was the priest of the most high God. And he blessed him, and said, Blessed be Abram of the most high God, possessor of heaven and earth: and blessed be the most high God, which hath delivered thine enemies into thy hand. And he gave him tithes of all. And the king of Sodom said unto Abram, Give me the persons, and take the goods to thyself. And Abram said to the king of Sodom, I have lifted up mine hand unto the Lord, the most high God, the possessor of heaven and earth; that I will not take from a thread even to a shoe-latchet, and that I will not take any thing that is thine, lest thou shouldest say, I have made Abram rich: save only that which the young men have eaten, and the portion of the men which went with me, Aner, Eshcol, and Mamre; let them take their portion.

After these things the word of the Lord came unto Abram in a vision, saying, Fear not, Abram: I am thy shield, and thy exceeding great reward. And Abram said, Lord God, what wilt thou give me, seeing I go childless, and the steward of my house is this Eliezer of Damascus? And Abram said, Behold, to me thou hast given no seed: and, lo, one born in my house is mine heir. And, behold, the word of the Lord came unto him, saying, This shall not be thine heir; but he that shall come forth out of thine own bowels shall be thine heir. And he brought him forth abroad, and said, Look now toward heaven, and tell the stars, if thou be able to number them: and he said unto him, So shall thy seed be. And he believed in the Lord; and he counted it to him for righteousness. And he said unto him, I am the Lord that brought thee out of Ur of the Chaldees, to give thee this land to inherit it. And he said, Lord God, whereby shall I know that I shall inherit it? And he said unto him, Take me an heifer of three years old, and a she-goat of three years old, and a ram of three years old, and a turtledove, and a young pigeon. And he took unto him all these, and divided them in the midst, and laid each piece one against another: but the birds divided he not. And when the fowls came down upon the carcases, Abram drove them away. And when the sun was going down, a deep sleep fell upon Abram; and, lo, an horror of great darkness fell upon him. And he said unto Abram, Know of a surety that thy seed shall be a stranger in a land that is not theirs, and shall serve them; and they shall afflict them four hundred years; and also that nation, whom they shall serve, will I judge: and afterward shall they come out with great substance. And thou shalt go to thy fathers in peace; thou shalt be buried in a good old age. But in the fourth generation they shall come hither again: for the iniquity of the Amorites is not yet full. And it came to pass, that, when the sun went down, and it was dark, behold a smoking furnace, and a burning lamp that passed between those pieces. In the same day the Lord made a covenant with Abram, saying, Unto thy seed have I given this land, from the river of Egypt unto the great river, the river Euphrates: the Kenites, and the Kenizzites, and the Kadmonites, and the Hittites, and the Perizzites, and the Rephaims, and the Amorites, and the Canaanites, and the Girgashites, and the Jebusites.

Now Sarai Abram's wife bare him no children: and she had an handmaid, an Egyptian, whose name was Hagar. And Sarai said unto Abram, Behold now, the Lord hath restrained me from bearing: I pray thee, go in unto my maid; it may be that I may obtain children by her. And Abram hearkened to the voice of Sarai. And Sarai Abram's wife took Hagar her maid the Egyptian, after Abram had dwelt ten years in the land of Canaan, and gave her to her husband Abram to be his wife.

And he went in unto Hagar, and she conceived: and when she saw that she had conceived, her mistress was despised in her eyes. And Sarai said unto Abram, My wrong be upon thee: I have given my maid into thy bosom; and when she saw that she had conceived, I was despised in her eyes: the Lord judge between me and thee. But Abram said unto Sarai, Behold, thy maid is in thine hand; do to her as it pleaseth thee. And when Sarai dealt hardly with her, she fled from her face.

And the angel of the Lord found her by a fountain of water in the wilderness, by the fountain in the way to Shur. And he said, Hagar, Sarai's maid, whence camest thou? and whither wilt thou go? And she said, I flee from the face of my mistress Sarai. And the angel of the Lord said unto her, Return to thy mistress, and submit thyself under her hands. And the angel of the Lord said unto her, I will multiply thy seed exceedingly, that it shall not be numbered for multitude. And the angel of the Lord said unto her, Behold, thou art with child, and shalt bear a son, and shalt call his name Ishmael; because the Lord hath heard thy affliction. And he will be a wild man; his hand will be against every man, and every man's hand against him; and he shall dwell in the presence of all his brethren. And she called the name of the Lord that spake unto her, Thou God seest me: for she said, Have I also here looked after him that seeth me? Wherefore the well was called Beer-lahairoi; behold, it is between Kadesh and Bered.

And Hagar bare Abram a son: and Abram called his son's name, which Hagar bare, Ishmael. And Abram was fourscore and six years old, when Hagar bare Ishmael to Abram.

And when Abram was ninety years old and nine, the Lord appeared to Abram, and said unto him, I am the Almighty God; walk before me, and be thou perfect. And I will make my covenant between me and thee, and will multiply thee exceedingly. And Abram fell on his face: and God talked with him, saying, As for me, behold, my covenant is with thee, and thou shalt be a father of many nations. Neither shall thy name any more be called Abram, but thy name shall be Abraham; for a father of many nations have I made thee. And I will make thee exceeding fruitful, and I will make nations of thee, and kings shall come out of thee. And I will establish my covenant between me and thee and thy seed after thee in their generations for an everlasting covenant, to be a God unto thee, and to thy seed after thee. And I will give unto thee, and to thy seed after thee, the land wherein thou art a stranger, all the land of Canaan, for an everlasting possession; and I will be their God.

And God said unto Abraham, Thou shalt keep my covenant therefore, thou, and thy seed after thee in their generations. This is my covenant, which ye shall keep, between me and you and thy seed after thee; Every man child among you shall be circumcised. And ye shall circumcise the flesh of your foreskin; and it shall be a token of the covenant betwixt me and you. And he that is eight days old shall be circumcised among you, every man child in your generations, he that is born in the house, or bought with money of any stranger, which is not of thy seed. He that is born in thy house, and he that is bought with thy money, must needs be circumcised: and my covenant shall be in your flesh for an everlasting covenant. And the uncircumcised man child whose flesh of his foreskin is not circumcised, that soul shall be cut off from his people; he hath broken my covenant.

And God said unto Abraham, As for Sarai thy wife, thou shalt not call her name Sarai, but Sarah shall her name be. And I will bless her, and give thee a son also of her: yea, I will bless her, and she shall be a mother of nations; kings of people shall be of her. Then Abraham fell upon his face, and laughed, and said in his heart, Shall a child be born unto him that is an hundred years old? and shall Sarah, that is ninety years old, bear? And Abraham said unto God, O that Ishmael might live before thee! And God said, Sarah thy wife shall bear thee a son indeed; and thou shalt call his name Isaac: and I will establish my covenant with him for an everlasting covenant, and with his seed after him. And as for Ishmael, I have heard thee: Behold, I have blessed him, and will make him fruitful, and will multiply him exceedingly; twelve princes shall he beget, and I will make him a great nation. But my covenant will I establish with Isaac, which Sarah shall bear unto thee at this set time in the next year. And he left off talking with him, and God went up from Abraham.

And Abraham took Ishmael his son, and all that were born in his house, and all that were bought with his money, every male among the men of Abraham's house; and circumcised the flesh of their foreskin in the selfsame day, as God had said unto him. And Abraham was ninety years old and nine, when he was circumcised in the flesh of his foreskin. And Ishmael his son was thirteen years old, when he was circumcised in the flesh of his foreskin. In the selfsame day was Abraham circumcised, and Ishmael his son. And all the men of his house, born in the house, and bought with money of the stranger, were circumcised with him.

And the Lord appeared unto him in the plains of Mamre: and he sat in the tent door in the heat of the day; and he lift up his eyes and looked, and, lo, three men stood by him: and when he saw them, he ran to meet them from the tent door, and bowed himself toward the ground, and said, My Lord, if now I have found favour in thy sight, pass not away, I pray thee, from thy servant: let a little water, I pray you, be fetched, and wash your feet, and rest yourselves under the tree: and I will fetch a morsel of bread, and comfort ye your hearts; after that ye shall pass on: for therefore are ye come to your servant. And they said, So do, as thou hast said. And Abraham hastened into the tent unto Sarah, and said, Make ready quickly three measures of fine meal, knead it, and make cakes upon the hearth. And Abraham ran unto the herd, and fetcht a calf tender and good, and gave it unto a young man; and he hasted to dress it. And he took butter, and milk, and the calf which he had dressed, and set it before them; and he stood by them under the tree, and they did eat.

And they said unto him, Where is Sarah thy wife? And he said, Behold, in the tent. And he said, I will certainly return unto thee according to the time of life; and, lo, Sarah thy wife shall have a son. And Sarah heard it in the tent door, which was behind him. Now Abraham and Sarah were old and well stricken in age; and it ceased to be with Sarah after the manner of women. Therefore Sarah laughed within herself, saying, After I am waxed old shall I have pleasure, my lord being old also? And the Lord said unto Abraham, Wherefore did Sarah laugh, saying, Shall I of a surety bear a child, which am old? Is any thing too hard for the Lord? At the time appointed I will return unto thee, according to the time of life, and Sarah shall have a son. Then Sarah denied, saying, I laughed not; for she was afraid. And he said, Nay; but thou didst laugh.

And the men rose up from thence, and looked toward Sodom: and Abraham went with them to bring them on the way. And the Lord said, Shall I hide from Abraham that thing which I do; seeing that Abraham shall surely become a great and mighty nation, and all the nations of the earth shall be blessed in him? For I know him, that he will command his children and his household after him, and they shall keep the way of the Lord, to do justice and judgment; that the Lord may bring upon Abraham that which he hath spoken of him. And the Lord said, Because the cry of Sodom and Gomorrah is great, and because their sin is very grievous; I will go down now, and see whether they have done altogether according to the cry of it, which is come unto me; and if not, I will know. And the men turned their faces from thence, and went toward Sodom: but Abraham stood yet before the Lord.

And Abraham drew near and said, Wilt thou also destroy the righteous with the wicked? Peradventure there be fifty righteous within the city: wilt thou also destroy and not spare the place for the fifty righteous that are therein? That be far from thee to do after this manner, to slay the righteous with the wicked: and that the righteous should be as the wicked, that be far from thee: Shall not the Judge of all the earth do right? And the Lord said, If I find in Sodom fifty righteous within the city, then I will spare all the place for their sakes. And

Abraham answered and said, Behold now, I have taken upon me to speak unto the Lord, which am but dust and ashes. Peradventure there shall lack five of the fifty righteous: wilt thou destroy all the city for lack of five? And he said, If I find there forty and five, I will not destroy it. And he spake unto him yet again, and said, Peradventure there shall be forty found there. And he said, I will not do it for forty's sake. And he said unto him, Oh let not the Lord be angry, and I will speak: Peradventure there shall thirty be found there. And he said, I will not do it, if I find thirty there. And he said, Behold now, I have taken upon me to speak unto the Lord: Peradventure there shall be twenty found there. And he said, I will not destroy it for twenty's sake. And he said, Oh let not the Lord be angry, and I will speak yet but this once: Peradventure ten shall be found there. And he said, I will not destroy it for ten's sake. And the Lord went his way, as soon as he had left communing with Abraham: and Abraham returned unto his place.

And there came two angels to Sodom at even: and Lot sat in the gate of Sodom: and Lot seeing them rose up to meet them; and he bowed himself with his face toward the ground; and he said, Behold now, my lords, turn in, I pray you, into your servant's house, and tarry all night, and wash your feet, and ye shall rise up early, and go on your ways. And they said, Nay; but we will abide in the street all night. And he pressed upon them greatly; and they turned in unto him, and entered into his house; and he made them a feast, and did bake unleavened bread, and they did eat.

But before they lay down, the men of the city, even the men of Sodom, compassed the house round, both old and young, all the people from every quarter: and they called unto Lot, and said unto him, Where are the men which came in to thee this night? bring them out unto us, that we may know them. And Lot went out at the door unto them, and shut the door after him, and said, I pray you, brethren, do not so wickedly. Behold now, I have two daughters which have not known man; let me, I pray you, bring them out unto you, and do ye to them as is good in your eyes: only unto these men do nothing; for therefore came they under the shadow of my roof. And they said, Stand back. And they said again, This one fellow came in to sojourn, and he will needs be a judge: now will we deal worse with thee, than with them. And they pressed sore upon the man, even Lot, and came near to break the door. But the men put forth their hand, and pulled Lot into the house to them, and shut to the door. And they smote the men that were at the door of the house with blindness, both small and great: so that they wearied themselves to find the door.

And the men said unto Lot, Hast thou here any besides? son in law, and thy sons, and thy daughters, and whatsoever thou hast in the city, bring them out of this place; for we will destroy this place, because the cry of them is waxen great before the face of the Lord; and the Lord hath sent us to destroy it. And Lot went out, and spake unto his sons in law, which married his daughters, and said, Up, get you out of this place; for the Lord will destroy this city. But he seemed as one that mocked unto his sons in law.

And when the morning arose, then the angels hastened Lot, saying, Arise, take thy wife, and thy two daughters, which are here; lest thou be consumed in the iniquity of the city. And while he lingered, the men laid hold upon his hand, and upon the hand of his wife, and upon the hand of his two daughters; the Lord being merciful unto him: and they brought him forth, and set him without the city.

And it came to pass, when they had brought them forth abroad, that he said, Escape for thy life; look not behind thee, neither stay thou in all the plain; escape to the mountain, lest thou be consumed. And Lot said unto them, Oh, not so, my Lord: behold now, thy servant hath found grace in thy sight, and thou hast magnified thy mercy, which thou hast shewed unto me in saving my life; and I cannot escape to the mountain, lest some evil take me, and I die: behold now, this city is near to flee unto, and it is a little one: Oh, let me escape thither (is it not a little one?) and my soul shall live. And he said unto him, See, I have accepted thee concerning this thing also, that I will not overthrow this city, for the which thou has spoken. Haste thee, escape thither; for I cannot do any thing till thou be come thither. Therefore the name of the city was called Zoar.

The sun was risen upon the earth when Lot entered into Zoar. Then the Lord rained upon Sodom and upon Gomorrah brimstone and fire from the Lord out of heaven; and he overthrew those cities, and all the plain, and all the inhabitants of the cities, and that which grew upon the ground.

But his wife looked back from behind him, and she became a pillar of salt.

And Abraham gat up early in the morning to the place where he stood before the Lord: and he looked toward Sodom and Gomorrah, and toward all the land of the plain, and beheld, and, lo, the smoke of the country went up as the smoke of a furnace.

And it came to pass, when God destroyed the cities of the plain, that God remembered Abraham, and sent Lot out of the midst of the overthrow, when he overthrew the cities in the which Lot dwelt.

And Lot went up out of Zoar, and dwelt in the mountain, and his two daughters with him; for he feared to dwell in Zoar: and he dwelt in a cave, he and his two daughters. And the firstborn said unto the younger, Our father is old, and there is not a man in the earth to come in unto us after the manner of all the earth: come, let us make our father drink wine, and we will lie with him, that we may preserve seed of our father. And they made their father drink wine that night: and the firstborn went in, and lay with her father; and he perceived not when she lay down, nor when she arose. And it came to pass on the morrow, that the firstborn said unto the younger, Behold, I lay yesternight with my father: let us make him drink wine this night also; and go thou in, and lie with him, that we may preserve seed of our father. And they made their father drink wine that night also: and the younger arose, and lay with him; and he perceived not when she lay down, nor when she arose. Thus were both the daughters of Lot with child by their father. And the firstborn bare a son, and called his name Moab: the same is the father of the Moabites unto this day. And the younger, she also bare a son, and called his name Ben-ammi: the same is the father of the children of Ammon unto this day.

And Abraham journeyed from thence toward the south country, and dwelled between Kadesh and Shur, and sojourned in Gerar. And Abraham said of Sarah his wife, She is my sister: and Abimelech king of Gerar sent, and took Sarah. But God came to Abimelech in a dream by night, and said to him, Behold, thou art but a dead man, for the woman which thou hast taken; for she is a man's wife. But Abimelech had not come near her: and he said, Lord, wilt thou slay also a righteous nation? Said he not unto me, She is my sister? and she, even she herself said, He is my brother: in the integrity of my heart and innocency of my hands have I done this. And God said unto him in a dream, Yea, I know that thou didst this in the integrity of thy heart; for I also withheld thee from sinning against me: therefore suffered I thee not to touch her. Now therefore restore the man his wife; for he is a prophet, and he shall pray for thee, and thou shalt live: and if thou restore her not, know thou that thou shalt surely die, thou, and all that are thine. Therefore Abimelech rose early in the morning, and called all his servants, and told all these things in their ears: and the men were sore afraid. Then Abimelech called Abraham, and said unto him, What hast thou done unto us? and what have I offended thee, that thou hast brought on me and on my kingdom a great sin? thou hast done deeds unto me that ought not to be done. And Abimelech said unto Abraham, What sawest thou, that thou hast done this thing? And Abraham said, Because I thought, Surely the fear of God is not in this place; and they will slay me for my wife's sake. And yet indeed she is my sister; she is the daughter of my father, but not the daughter of my mother; and she became my wife. And it came to pass, when God caused me to wander from my father's house, that I said unto her, This is thy kindness which thou shalt shew unto me; at every place whither we shall come, say of me, He is my brother. And Abimelech took sheep, and oxen, and menservants, and womenservants, and gave them unto Abraham, and restored him Sarah his wife. And Abimelech said, Behold, my land is before thee: dwell where it pleaseth thee. And unto Sarah he said, Behold, I have given thy brother a thousand pieces of silver: behold, he is to thee a covering of the eyes, unto all that are with thee, and with all other: thus she was reproved.

So Abraham prayed unto God: and God healed Abimelech, and his wife, and his maidservants; and they bare children. For the Lord had fast closed up all the wombs of the house of Abimelech, because of Sarah Abraham's wife.

And the Lord visited Sarah as he had said, and the Lord did unto Sarah as he had spoken. For Sarah conceived, and bare Abraham a son in his old age, at the set time of which God had spoken to him. And Abraham called the name of his son that was born unto him, whom Sarah bare to him, Isaac. And Abraham circumcised his son Isaac being eight days old, as God had commanded him. And Abraham was an hundred years old, when his son Isaac was born unto him.

And Sarah said, God hath made me to laugh, so that all that hear will laugh with me. And she said, Who would have said unto Abraham, that Sarah should have given children suck? for

I have born him a son in his old age. And the child grew, and was weaned: and Abraham made a great feast the same day that Isaac was weaned.

And Sarah saw the son of Hagar the Egyptian, which she had born unto Abraham, mocking. Wherefore she said unto Abraham, Cast out this bondwoman and her son: for the son of this bondwoman shall not be heir with my son, even with Isaac. And the thing was very grievous in Abraham's sight because of his son.

And God said unto Abraham, Let it not be grievous in thy sight because of the lad, and because of thy bondwoman; in all that Sarah hath said unto thee, hearken unto her voice; for in Isaac shall thy seed be called. And also of the son of the bond-woman will I make a nation, because he is thy seed. And Abraham rose up early in the morning, and took bread, and a bottle of water, and gave it unto Hagar, putting it on her shoulder, and the child, and sent her away: and she departed, and wandered in the wilderness of Beer-sheba. And the water was spent in the bottle, and she cast the child under one of the shrubs. And she went, and sat her down over against him a good way off, as it were a bowshot: for she said, Let me not see the death of the child. And she sat over against him, and lifted up her voice, and wept. And God heard the voice of the lad; and the angel of God called to Hagar out of heaven, and said unto her, What aileth thee, Hagar? fear not; for God hath heard the voice of the lad where he is. Arise, lift up the lad, and hold him in thine hand; for I will make him a great nation. And God opened her eyes, and she saw a well of water; and she went, and filled the bottle with water, and gave the lad drink. And God was with the lad; and he grew, and dwelt in the wilderness, and became an archer. And he dwelt in the wilderness of Paran: and his mother took him a wife out of the land of Egypt.

And it came to pass at that time, that Abimelech and Phichol the chief captain of his host spake unto Abraham, saying, God is with thee in all that thou doest: now therefore swear unto me here by God that thou wilt not deal falsely with me, nor with my son, nor with my son's son: but according to the kindness that I have done unto thee, thou shalt do unto me, and to the land wherein thou hast sojourned. And Abraham said, I will swear.

And Abraham reproved Abimelech because of a well of water, which Abimelech's servants had violently taken away. And Abimelech said, I know not who hath done this thing: neither didst thou tell me, neither yet heard I of it, but to day. And Abraham took sheep and oxen, and gave them unto Abimelech; and both of them made a covenant. And Abraham set seven ewe lambs of the flock by themselves. And Abimelech said unto Abraham, What mean these seven ewe lambs which thou hast set by themselves? And he said, For these seven ewe lambs shalt thou take of my hand, that they may be a witness unto me, that I have digged this well. Wherefore he called that place Beer-sheba; because there they sware both of them. Thus they made a covenant at Beer-sheba; then Abimelech rose up, and Phichol the chief captain of his host, and they returned into the land of the Philistines.

And Abraham planted a grove in Beer-sheba, and called there on the name of the Lord, the everlasting God. And Abraham sojourned in the Philistines' land many days.

And it came to pass after these things, that God did tempt Abraham, and said unto him, Abraham: and he said, Behold, here I am. And he said, Take now thy son, thine only son Isaac, whom thou lovest, and get thee into the land of Moriah; and offer him there for a burnt offering upon one of the mountains which I will tell thee of.

And Abraham rose up early in the morning, and saddled his ass, and took two of his young men with him, and Isaac his son, and clave the wood for the burnt offering, and rose up, and went unto the place of which God had told him. Then on the third day Abraham lifted up his eyes, and saw the place afar off. And Abraham said unto his young men, Abide ye here with the ass; and I and the lad will go yonder and worship, and come again to you. And Abraham took the wood of the burnt offering, and laid it upon Isaac his son; and he took the fire in his hand, and a knife; and they went both of them together. And Isaac spake unto Abraham his father, and said, My father: and he said, Here am I, my son. And he said, Behold the fire and the wood: but where is the lamb for a burnt offering? And Abraham said, My son, God will provide himself a lamb for a burnt offering: so they went both of them together. And they came to the place which God had told him of; and Abraham built an altar there, and laid the wood in order, and bound Isaac his son, and laid him on the altar upon the wood. And Abraham stretched forth his hand, and took the knife to slay his son. And the angel of the Lord called unto him out of heaven, and said, Abraham, Abraham: and he said, Here am I. And he said, Lay not thine hand upon the lad, neither do thou any thing unto him: for now I know that thou fearest God, seeing thou hast not withheld thy son, thine only son from me. And Abraham lifted up his eyes, and looked, and behold behind him a ram caught in a thicket by his horns: and Abraham went and took the ram, and offered him up for a burnt offering in the stead of his son. And Abraham called the name of that place Jehovah-jireh: as it is said to this day, In the mount of the Lord it shall be seen.

And the angel of the Lord called unto Abraham out of heaven the second time, and said, By myself have I sworn, saith the Lord, for because thou hast done this thing, and hast not with-held thy son, thine only son: that in blessing I will bless thee, and in multiplying I will multiply thy seed as the stars of the heaven, and as the sand which is upon the sea shore; and thy seed shall possess the gate of his enemies; and in thy seed shall all the nations of the earth be blessed; because thou hast obeyed my voice. So Abraham returned unto his young men, and they rose up and went together to Beer-sheba; and Abraham dwelt at Beer-sheba. . . .

And Sarah was an hundred and seven and twenty years old: these were the years of the life of Sarah. And Sarah died in Kirjath-arba; the same is Hebron in the land of Canaan: and Abraham came to mourn for Sarah, and to weep for her.

And Abraham stood up from before his dead, and spake unto the sons of Heth, saying, I am a stranger and a sojourner with you: give me a possession of a buryingplace with you, that I may bury my dead out of my sight. And the children of Heth answered Abraham, saying unto him, Hear us, my lord: thou art a mighty prince among us: in the choice of our sepulchres bury thy dead; none of us shall withhold from thee his sepulchre, but that thou mayest bury thy dead. And Abraham stood up, and bowed himself to the people of the land, even to the children of Heth. And he communed with them, saying, If it be your mind that I should bury my dead out of my sight; hear me, and intreat for me to Ephron the son of Zohar, that he may give me the cave of Machpelah, which he hath, which is in the end of his field; for as much money as it is worth he shall give it me for a possession of a buryingplace amongst you. And Ephron dwelt among the children of Heth: and Ephron the Hittite answered Abraham in the audience of the children of Heth, even of all that went in at the gate of his city, saying, Nay, my lord, hear me: the field give I thee, and the cave that is therein, I give it thee; in the presence of the sons of my people give I it thee: bury thy dead. And Abraham bowed down himself before the people of the land. And he spake unto Ephron in the audience of the people of the land, saying, But if thou wilt give it, I pray thee, hear me: I will give thee money for the field; take it of me, and I will bury my dead there. And Ephron answered Abraham, saying unto him, My lord, hearken unto me: the land is worth four hundred shekels of silver; what is that betwixt me and thee? bury therefore thy dead. And Abraham hearkened unto Ephron; and Abraham weighed to Ephron the silver, which he had named in the audience of the sons of Heth, four hundred shekels of silver, current money with the merchant.

And the field of Ephron, which was in Machpelah, which was before Mamre, the field, and the cave which was therein, and all the trees that were in the field, that were in all the borders round about, were made sure unto Abraham for a possession in the presence of the children of Heth, before all that went in at the gate of his city. And after this, Abraham buried Sarah his wife in the cave of the field of Machpelah before Mamre: the same is Hebron in the land of Canaan. And the field, and the cave that is therein, were made sure unto Abraham for a possession of a buryingplace by the sons of Heth.

SIGMUND FREUD was born in 1856 in Freiburg, Moravia. His father was a wool merchant who moved the family to Vienna in 1860. There, at the University of Vienna, Freud studied medicine and carried out research on neurological disorders such as Meynert's Amentia (hallucinatory psychosis) and the effects of cocaine on the body. Despite Freud's achievements as the founder of psychoanalysis, he recalled modestly of his student days: ". . . I was compelled to make the discovery that the peculiarities and limitations of my gifts denied me all success in many of the departments of science. . . ." Freud joined the staff of Theodor Meynert's psychiatric clinic in 1882 and was appointed a lecturer in neuropathology at Vienna from 1883–85. He married and set up a practice, meanwhile writing and conducting psychoanalytic research. Freud's works include *The Interpretation of Dreams* (1900), *The Origin and Development of Psychoanalysis* (1910), *Totem and Taboo* (1913), and *Civilization and Its Discontents* (1930). Freud died in London in 1939.

A selection from *Civilization and Its Discontents,* translated by James Strachey. Publisher: W. W. Norton & Co., Inc., 1962.

Civilization
and Its Discontents

It is impossible to escape the impression that people commonly use false standards of measurement—that they seek power, success and wealth for themselves and admire them in others, and that they underestimate what is of true value in life. And yet, in making any general judgment of this sort, we are in danger of forgetting how variegated the human world and its mental life are. There are a few men from whom their contemporaries do not withhold admiration, although their greatness rests on attributes and achievements which are completely foreign to the aims and ideals of the multitude. One might easily be inclined to suppose that it is after all only a minority which appreciates these great men, while the large majority cares nothing for them. But things are probably not as simple as that, thanks to the discrepancies between people's thoughts and their actions, and to the diversity of their wishful impulses.

One of these exceptional few calls himself my friend in his letters to me. I had sent him my small book that treats religion as an illusion,[1] and he answered that he entirely agreed with my judgment upon religion, but that he was sorry I had not properly appreciated the true source of religious sentiments. This, he says, consists in a peculiar feeling, which he himself is never

---

[1] [*The Future of an Illusion.*]

without, which he finds confirmed by many others, and which he may suppose is present in millions of people. It is a feeling which he would like to call a sensation of "eternity," a feeling as of something limitless, unbounded—as it were, "oceanic." This feeling, he adds, is a purely subjective fact, not an article of faith; it brings with it no assurance of personal immortality, but it is the source of the religious energy which is seized upon by the various Churches and religious systems, directed by them into particular channels, and doubtless also exhausted by them. One may, he thinks, rightly call oneself religious on the ground of this oceanic feeling alone, even if one rejects every belief and every illusion.

The views expressed by the friend whom I so much honor, and who himself once praised the magic of illusion in a poem, caused me no small difficulty. I cannot discover this "oceanic" feeling in myself. It is not easy to deal scientifically with feelings. One can attempt to describe their physiological signs. Where this is not possible—and I am afraid that the oceanic feeling too will defy this kind of characterization—nothing remains but to fall back on the ideational content which is most readily associated with the feeling. If I have understood my friend rightly, he means the same thing by it as the consolation offered by an original and somewhat eccentric dramatist to his hero who is facing a self-inflicted death. "We cannot fall out of this world." That is to say, it is a feeling of an indissoluble bond, of being one with the external world as a whole. I may remark that to me this seems something rather in the nature of an intellectual perception, which is not, it is true, without an ac- companying feeling-tone, but only such as would be present with any other act of thought of equal range. From my own experience I could not convince myself of the primary nature of such a feeling. But this gives me no right to deny that it does in fact occur in other people. The only question is whether it is being correctly interpreted and whether it ought to be regarded as the *fons et origo*[2] of the whole need for religion.

I have nothing to suggest which could have a decisive influence on the solution of this problem. The idea of men's receiving an imitation of their connection with the world around them through an immediate feeling which is from the outset directed to that purpose sounds so strange and fits in so badly with the fabric of our psychology that one is justified in attempting to discover a psychoanalytic . . . explanation of such a feeling. The following line of thought suggests itself. Normally, there is nothing of which we are more certain than the feeling of our self, of our own ego. This ego appears to us as something autonomous and unitary, marked off distinctly from everything else . . . .

The adult's ego-feeling cannot have been the same from the beginning. It must have gone through a process of development, which cannot, of course, be demonstrated but which admits of being constructed with a fair degree of probability. An infant at the breast does not as yet distinguish his ego from the external world as the source of the sensations flowing in upon him. He gradually learns to do so, in response to various promptings. He must be very strongly impressed by the fact that some sources of excitation, which he will later recognize as his own bodily organs, can provide him with sensations at any moment, whereas other sources evade him from time to time—among them what he desires most of all, his mother's breast—and only reappear as a result of his screaming for help. In this way there is for the first time set over against the ego an "object," in the form of something which exists "outside" and which is only forced to appear by a special action. . . .

In this way, then, the ego detaches itself from the external world. Or, to put it more correctly, originally the ego includes everything, later it separates off an external world from itself.

---

[2] [source and origin.]

Our present ego-feeling is, therefore, only a shrunken residue of a much more inclusive—indeed, an all-embracing—feeling which corresponded to a more intimate bond between the ego and the world about it. If we may assume that there are many people in whose mental life this primary ego-feeling has persisted to a greater or less degree, it would exist in them side by side with the narrower and more sharply demarcated ego-feeling of maturity, like a kind of counterpart to it. In that case, the ideational contents appropriate to it would be precisely those of limitlessness and of a bond with the universe—the same ideas with which my friend elucidated the "oceanic" feeling.

But have we a right to assume the survival of something that was originally there, alongside of what was later derived from it? Undoubtedly. There is nothing strange in such a phenomenon, whether in the mental field or elsewhere. In the animal kingdom we hold to the view that the most highly developed species have proceeded from the lowest; and yet we find all the simple forms still in existence today. The race of the great saurians is extinct and has made way for the mammals; but a true representative of it, the crocodile, still lives among us. This analogy may be too remote, and it is also weakened by the circumstance that the lower species which survive are for the most part not the true ancestors of the present-day more highly developed species. As a rule the intermediate links have died out and are known to us only through reconstruction. In the realm of the mind, on the other hand, what is primitive is so commonly preserved alongside of the transformed version which has arisen from it that it is unnecessary to give instances as evidence. When this happens it is usually in consequence of a divergence in development: one portion (in the quantitative sense) of an attitude or instinctual impulse has remained un-altered, while another portion has undergone further develop-ment. . . .

Thus we are perfectly willing to acknowledge that the "oceanic" feeling exists in many people, and we are inclined to trace it back to an early phase of ego-feeling. The further question then arises, what claim this feeling has to be regarded as the source of religious needs.

To me the claim does not seem compelling. After all, a feeling can only be a source of energy if it is itself the expression of a strong need. The derivation of religious needs from the infant's helplessness and the longing for the father aroused by it seems to me incontrovertible, especially since the feeling is not simply prolonged from childhood days, but is permanently sustained by fear of the superior power of Fate. I cannot think of any need in childhood as strong as the need for a father's protection. Thus the part played by the oceanic feeling . . . is ousted from a place in the foreground. The origin of the religious attitude can be traced back in clear outlines as far as the feeling of infantile helplessness. There may be something further behind that, but for the present it is wrapped in obscurity.

I can imagine that the oceanic feeling became connected with religion later on. The "oneness with the universe" which constitutes its ideational content sounds like a first attempt at a religious consolation, as though it were another way of disclaiming the danger which the ego recognizes as threatening it from the external world. Let me admit once more that it is very difficult for me to work with these almost intangible quantities. Another friend of mine, whose insatiable craving for knowledge has led him to make the most unusual experiments and has ended by giving him encyclopedic knowledge, has assured me that through the practices of yoga, by withdrawing from the world, by fixing the attention on bodily functions and by peculiar methods of breathing, one can in fact evoke new sensations and coenaesthesias in oneself, which he regards as regressions to primordial states of mind which have long ago been overlaid. He sees in them a physiological basis, as it were, of much of the wisdom of mysticism. It would not be hard to find connections here with a number of obscure modifications of mental life, such as trances and ecstasies. . . .

In my *Future of an Illusion* I was concerned much less with the deepest sources of the religious feeling than with what the common man understands by his religion—with the system of doctrines and promises which on the one hand explains to him the riddles of this world with enviable completeness, and, on the other, assures him that a careful Providence will watch over his life and will compensate him in a future existence for any frustrations he suffers here. The common man cannot imagine this Providence otherwise than in the figure of an enormously exalted father. Only such a being can understand the needs of the children of men and be softened by their prayers and placated by the signs of their remorse. The whole thing is so patently infantile, so foreign to reality, that to anyone with a friendly attitude to humanity it is painful to think that the great majority of mortals will never be able to rise above this view of life. It is still more humiliating to discover how large a number of people living today, who cannot but see that this religion is not tenable, nevertheless try to defend it piece by piece in a series of pitiful rearguard actions. One would like to mix among the ranks of the believers in order to meet these philosophers, who think they can rescue the God of religion by replacing him by an impersonal, shadowy and abstract principle, and to address them with the warning words: "Thou shalt not take the name of the Lord thy God in vain!" And if some of the great men of the past acted in the same way, no appeal can be made to their example: we know why they were obliged to.

Let us return to the common man and to his religion—the only religion which ought to bear that name. The first thing we think of is the well-known saying of one of our great poets and thinkers concerning the relation of religion to art and science:

> Wer Wissenschaft und Kunst besitzt, hat auch Religion;
> Wer jene beide nicht besitzt, der habe Religion![3]

---

[3] ["He who possesses science and art also has religion; but he who possesses

This saying on the one hand draws an antithesis between religion and the two highest achievements of man and on the other, asserts that, as regards their value in life, those achievements and religion can represent or replace each other. If we also set out to deprive the common man [who has neither science nor art] of his religion, we shall clearly not have the poet's authority on our side. We will choose a particular path to bring us nearer an appreciation of his words. Life, as we find it, is too hard for us; it brings us too many pains, disappointments and impossible tasks. In order to bear it we cannot dispense with palliative measures. "We cannot do without auxiliary constructions," as Theodor Fontane tells us. There are perhaps three such measures: powerful deflections, which cause us to make light of our misery; substitutive satisfactions, which diminish it; and intoxicating substances, which make us insensitive to it. Something of the kind is indispensable. Voltaire has deflections in mind when he ends *Candide* with the advice to cultivate one's garden; and scientific activity is a deflection of this kind, too. The substitutive satisfactions, as offered by art, are illusions in contrast with reality, but they are none the less psychically effective, thanks to the role which fantasy has assumed in mental life. The intoxicating substances influence our body and alter its chemistry. It is no simple matter to see where religion has its place in this series. We must look further afield.

The question of the purpose of human life has been raised countless times; it has never yet received a satisfactory answer and perhaps does not admit of one. Some of those who have asked it have added that if it should turn out that life has *no* purpose, it would lose all value for them. But this threat alters nothing. It looks, on the contrary, as though one had a right to dismiss the question, for it seems to derive from the human

---

neither of those two, let him have religion."—Goethe, *Zahme Xenien IX.*
—TRANS.]

presumptuousness, many other manifestations of which are already familiar to us. Nobody talks about the purpose of the life of animals, unless, perhaps, it may be supposed to lie in being of service to man. But this view is not tenable either, for there are many animals of which men can make nothing, except to describe, classify, and study them; and innumerable species of animals have escaped even this use, since they existed and became extinct before man set eyes on them. Once again, only religion can answer the question of the purpose of life. One can hardly be wrong in concluding that the idea of life having a purpose stands and falls with the religious system.

We will therefore turn to the less ambitious question of what men themselves show by their behavior to be the purpose and intention of their lives. What do they demand of life and wish to achieve in it? The answer to this can hardly be in doubt. They strive after happiness; they want to become happy and to remain so. This endeavor has two sides, a positive and a negative aim. It aims, on the one hand, at an absence of pain and unpleasure, and, on the other, at the experiencing of strong feelings of pleasure. In its narrower sense the word "happiness" only relates to the last. In conformity with this dichotomy in his aims, man's activity develops in two directions, according as it seeks to realize—in the main, or even exclusively—the one or the other of these aims.

As we see, what decides the purpose of life is simply the program of the pleasure principle.[4] This principle dominates the operation of the mental apparatus from the start. There can be no doubt about its efficacy, and yet its program is at loggerheads with the whole world, with the macrocosm as much as with the microcosm. There is no possibility at all of its being carried through; all the regulations of the universe run counter to it.

---

[4] [The principle that all mental activity is directed towards avoiding unpleasure and attaining pleasure.]

One feels inclined to say that the intention that man should be "happy" is not included in the plan of "Creation." What we call happiness in the strictest sense comes from the (preferably sudden) satisfaction of needs which have been dammed up to a high degree, and it is from its nature only possible as an episodic phenomenon. When any situation that is desired by the pleasure principle is prolonged, it only produces a feeling of mild contentment. We are so made that we can derive intense enjoyment only from a contrast and very little from a state of things. Thus our possibilities of happiness are already restricted by our constitution. Unhappiness is much less difficult to experience. We are threatened with suffering from three directions: from our own body, which is doomed to decay and dissolution and which cannot even do without pain and anxiety as warning signals; from the external world, which may rage against us with overwhelming and merciless forces of destruction; and finally from our relations to other men. The suffering which comes from this last source is perhaps more painful to us than any other. We tend to regard it as a kind of gratuitous addition, although it cannot be any less fatefully inevitable than the suffering which comes from elsewhere. . . .

The program of becoming happy, which the pleasure principle imposes on us, cannot be fulfilled; yet we must not—indeed, we cannot—give up our efforts to bring it nearer to fulfillment by some means or other. Very different paths may be taken in that direction, and we may give priority either to the positive aspect of the aim, that of gaining pleasure, or to its negative one, that of avoiding unpleasure. By none of these paths can we attain all that we desire. . . . There is no golden rule which applies to everyone: every man must find out for himself in what particular fashion he can be saved. All kinds of different factors will operate to direct his choice. It is a question of how much real satisfaction he can expect to get from the external world, how far he is led to make himself independent of it,

and, finally, how much strength he feels he has for altering the
world to suit his wishes. In this, his psychical constitution will
play a decisive part, irrespectively of the external circumstances.
The man who is predominantly erotic will give first preference
to his emotional relationships to other people; the narcissistic
man, who inclines to be self-sufficient, will seek his main satisfactions in his internal mental processes; the man of action will
never give up the external world on which he can try out his
strength. As regards the second of these types, the nature of his
talents and the amount of instinctual sublimation open to him
will decide where he shall locate his interests. Any choice that
is pushed to an extreme will be penalized by exposing the individual to the dangers which arise if a technique of living that
has been chosen as an exclusive one should prove inadequate.
Just as a cautious businessman avoids tying up all his capital
in one concern, so, perhaps, worldly wisdom will advise us not
to look for the whole of our satisfaction from a single aspiration.
Its success is never certain, for that depends on the convergence
of many factors, perhaps on none more than on the capacity of
the psychical constitution to adapt its function to the environment and then to exploit that environment for a yield of pleasure.
A person who is born with a specially unfavorable instinctual
constitution, and who has not properly undergone the transfor-
mation and rearrangement of his libidinal[5] components which
is indispensable for later achievements, will find it hard to obtain
happiness from his external situation, especially if he is faced
with tasks of some difficulty. As a last technique of living, which
will at least bring him substitutive satisfactions, he is offered
that of a flight into neurotic illness—a flight which he usually
accomplishes when he is still young. The man who sees his
pursuit of happiness come to nothing in later years can still find
consolation in the yield of pleasure of chronic intoxication; or

---

[5] [Emotional or psychic energy originally directed toward sexual love.]

he can embark on the desperate attempt at rebellion seen in a psychosis.

Religion restricts this play of choice and adaptation, since it imposes equally on everyone its own path to the acquisition of happiness and protection from suffering. Its technique consists in depressing the value of life and distorting the picture of the real world in a delusional manner—which presupposes an intimidation of the intelligence. At this price, by forcibly fixing them in a state of psychical infantilism and by drawing them into a mass-delusion, religion succeeds in sparing many people an individual neurosis. But hardly anything more. There are, as we have said, many paths which *may* lead to such happiness as is attainable by men, but there is none which does so for certain. Even religion cannot keep its promise. If the believer finally sees himself obliged to speak of God's "inscrutable decrees," he is admitting that all that is left to him as a last possible consolation and source of pleasure in his suffering is an unconditional submission. And if he is prepared for that, he could probably have spared himself the *détour* he has made.

Our inquiry concerning happiness has not so far taught us much that is not already common knowledge. And even if we proceed from it to the problem of why it is so hard for men to be happy, there seems no greater prospect of learning anything new. We have given the answer already by pointing to the three sources from which our suffering comes: the superior power of nature, the feebleness of our own bodies, and the inadequacy of the regulations which adjust the mutual relationships of human beings in the family, the state, and society. In regard to the first two sources, our judgment cannot hesitate long. It forces us to acknowledge those sources of suffering and to submit to the inevitable. We shall never completely master nature; and our bodily organism, itself a part of that nature, will always remain a transient structure with a limited capacity for adaptation and achievement. This recognition does not have a paralyzing effect. On the contrary, it points the direction for our activity. If we cannot remove all suffering, we can remove some, and we can mitigate some: the experience of many thousands of years has convinced us of that. As regards the third source, the social source of suffering, our attitude is a different one. We do not admit it at all; we cannot see why the regulations made by ourselves should not, on the contrary, be a protection and a benefit for every one of us. And yet, when we consider how unsuccessful we have been in precisely this field of prevention of suffering, a suspicion dawns on us that here, too, a piece of unconquerable nature may lie behind—this time a piece of our own psychical constitution.

When we start considering this possibility, we come upon a contention which is so astonishing that we must dwell upon it. This contention holds that what we call our civilization is largely responsible for our misery, and that we should be much happier if we gave it up and returned to primitive conditions. I call this contention astonishing because, in whatever way we may define the concept of civilization, it is a certain fact that all the things with which we seek to protect ourselves against the threats that emanate from the sources of suffering are part of that very civilization.

How has it happened that so many people have come to take up this strange attitude of hostility to civilization? I believe that the basis of it was a deep and long-standing dissatisfaction with the then existing state of civilization and that on that basis a condemnation of it was built up, occasioned by certain specific historical events. I think I know what the last and the last but one of those occasions were. I am not learned enough to trace the chain of them far back enough in the history of the human species; but a factor of this kind hostile to civilization must already have been at work in the victory of Christendom over the heathen religions. For it was very closely related to the low estimation put upon earthly life by the Christian doctrine. The last but one of these occasions was when the progress of voyages of discovery led to contact with primitive peoples and races. In consequence of insufficient observation and a mistaken view of their manners and customs, they appeared to Europeans to be leading a simple, happy life with few wants, a life such as was unattainable by their visitors with their superior civilization. Later experience has corrected some of those judgments. In many cases the observers had wrongly attributed to the absence of complicated cultural demands what was in fact due to the bounty of nature and the ease with which the major human needs were satisfied. The last occasion is especially familiar to us. It arose when people came to know about the mechanism of the neuroses, which threaten to undermine the modicum of happiness enjoyed by civilized men. It was discovered that a person becomes neurotic because he cannot tolerate the amount of frustration which society imposes on him in the service of its cultural ideals, and it was inferred from this that the abolition or reduction of those demands would result in a return to possibilities of happiness.

There is also an added factor of disappointment. During the last few generations mankind has made an extraordinary advance in the natural sciences and in their technical application and has established his control over nature in a way never before imagined. The single steps of this advance are common knowledge and it is unnecessary to enumerate them. Men are proud of those achievements, and have a right to be. But they seem to have observed that this newly-won power over space and time, this subjugation of the forces of nature, which is the fulfillment of a longing that goes back thousands of years, has not increased the amount of pleasurable satisfaction which they may expect from life and has not made them feel happier. From the recognition of this fact we ought to be content to conclude that power over nature is not the *only* precondition of human happiness, just as it is not the *only* goal of cultural endeavor; we ought not to infer from it that technical progress is without value for the economics of our happiness. One would like to ask: is there, then, no positive gain in pleasure, no unequivocal increase in my feeling of happiness, if I can, as often as I please, hear the voice of a child of mine who is living hundreds of miles away or if I can learn in the shortest possible time after a friend has reached his destination that he has come through the long and difficult voyage unharmed? Does it mean nothing that medicine has succeeded in enormously reducing infant mortality and the danger of infection for women in childbirth, and, indeed, in considerably lengthening the average life of a civilized man? And there is a long list that might be added to benefits of this kind which we owe to the much-despised era of scientific and technical advances. But here the voice of pessimistic criticism makes itself heard and warns us that most of these satisfactions follow the model of the "cheap enjoyment" extolled in the anecdote—the enjoyment obtained by putting a bare leg from under the bedclothes on a cold winter night and drawing it in again. If there had been no railway to conquer distances, my child would never have left his native town and I should need no telephone to hear his voice; if traveling across the ocean by ship had not been introduced, my friend would not have embarked on his sea-voyage and I should not need a cable to relieve my anxiety about him. What is the use of reducing infantile mortality when it is precisely that reduction which imposes the greatest restraint on us in the begetting of children, so that, taken all round, we nevertheless rear no more children than in the days before the reign of hygiene, while at the same time we have created difficult conditions for our sexual life in marriage, and have probably worked against the beneficial effects of natural selection? And, finally, what good to us is a long life if it is difficult and barren of joys, and if it is so full of misery that we can only welcome death as a deliverer?

*   *   *

[An important feature of civilization is] the manner in which
the relationships of men to one another, their social relationships,
are regulated—relationships which affect a person as a neighbor,
as a source of help, as another person's sexual object, as a member
of a family and of a State. Here it is especially difficult to keep
clear of particular ideal demands and to see what is civilized in
general. Perhaps we may begin by explaining that the element
of civilization enters on the scene with the first attempt to regulate these social relationships. If the attempt were not made,
the relationships would be subject to the arbitrary will of the
individual: that is to say, the physically stronger man would
decide them in the sense of his own interests and instinctual
impulses. Nothing would be changed in this if this stronger
man should in his turn meet someone even stronger than he.
Human life in common is only made possible when a majority
comes together which is stronger than any separate individual
and which remains united against all separate individuals. The
power of this community is then set up as "right" in opposition
to the power of the individual, which is condemned as "brute
force." This replacement of the power of the individual by the
power of a community constitutes the decisive step of civilization. The essence of it lies in the fact that the members of
the community restrict themselves in their possibilities of satisfaction, whereas the individual knew no such restrictions. The
first requisite of civilization, therefore, is that of justice—that
is, the assurance that a law once made will not be broken in
favor of an individual. This implies nothing as to the ethical
value of such a law. The further course of cultural development
seems to tend towards making the law no longer an expression
of the will of a small community—a caste or a stratum of the
population or a racial group—which, in its turn, behaves like
a violent individual towards other, and perhaps more numerous,

collections of people. The final outcome should be a rule of law to which all—except those who are not capable of entering a community—have contributed by a sacrifice of their instincts, and which leaves no one—again with the same exception—at the mercy of brute force.

The liberty of the individual is no gift of civilization. It was greatest before there was any civilization, though then, it is true, it had for the most part no value, since the individual was scarcely in a position to defend it. The development of civilization imposes restrictions on it, and justice demands that no one shall escape those restrictions. What makes itself felt in a human community as a desire for freedom may be their revolt against some existing injustice, and so may prove favorable to a further development of civilization; it may remain compatible with civilization. But it may also spring from the remains of their original personality, which is still untamed by civilization and may thus become the basis in them of hostility to civilization. The urge for freedom, therefore, is directed against particular forms and demands of civilization or against civilization altogether. It does not seem as though any influence could induce a man to change his nature into a termite's. No doubt he will always defend his claim to individual liberty against the will of the group. A good part of the struggles of mankind center round the single task of finding an expedient accommodation—one, that is, that will bring happiness—between this claim of the individual and the cultural claims of the group; and one of the problems that touches the fate of humanity is whether such an accommodation can be reached by means of some particular form of civilization or whether this conflict is irreconcilable.

*   *   *

The clue may be supplied by one of the ideal demands, as we have called them, of civilized society. It runs: "Thou shalt love thy neighbor as thyself." It is known throughout the world and is undoubtedly older than Christianity, which puts it forward as its proudest claim. Yet it is certainly not very old; even in historical times it was still strange to mankind. Let us adopt a naive attitude towards it, as though we were hearing it for the first time; we shall be unable then to suppress a feeling of surprise and bewilderment. Why should we do it? What good will it do us? But, above all, how shall we achieve it? How can it be possible? My love is something valuable to me which I ought not to throw away without reflection. It imposes duties on me for whose fulfillment I must be ready to make sacrifices. If I love someone, he must deserve it in some way. (I leave out of account the use he may be to me, and also his possible significance for me as a sexual object, for neither of these two kinds of relationship comes into question where the precept to love my neighbor is concerned.) He deserves it if he is so like me in important ways that I can love myself in him; and he deserves it if he is so much more perfect than myself that I can love my ideal of my own self in him. Again, I have to love him if he is my friend's son, since the pain my friend would feel if any harm came to him would be my pain too—I should have to share it. But if he is a stranger to me and if he cannot attract me by any worth of his own or any significance that he may already have acquired for my emotional life, it will be hard for me to love him. Indeed, I should be wrong to do so, for my love is valued by all my own people as a sign of my preferring them, and it is an injustice to them if I put a stranger on a par with them. But if I am to love him (with this universal love) merely because he, too, is an inhabitant of this earth, like an insect, an earthworm, or a grass-snake, then I fear that only a small modicum of my love will fall to his share—not by any possibility as much as, by the judgment of my reason, I am entitled to retain for myself. What is the point of a precept enunciated with so much solemnity if its fulfillment cannot be recommended as reasonable?

On closer inspection, I find still further difficulties. Not merely is this stranger in general unworthy of my love; I must honestly confess that he has more claim to my hostility and even my hatred. He seems not to have the least trace of love for me, and shows me not the slightest consideration. If it will do him any good he has no hesitation in injuring me, nor does he ask himself whether the amount of advantage he gains bears any proportion to the extent of the harm he does to me. Indeed, he need not even obtain an advantage; if he can satisfy any sort of desire by it, he thinks nothing of jeering at me, insulting me, slandering me, and showing his superior power; and the more secure he feels and the more helpless I am, the more certainly I can expect him to behave like this to me. If he behaves differently, if he shows me consideration and forbearance as a stranger, I am ready to treat him in the same way, in any case and quite apart from any precept. Indeed, if this grandiose commandment had run "Love thy neighbor as thy neighbor loves thee," I should not take exception to it. And there is a second commandment, which seems to me even more incomprehensible and arouses still stronger opposition in me. It is "Love thine enemies." If I think it over, however, I see that I am wrong in treating it as a greater imposition. At bottom it is the same thing. . . .

Now it is very probable that my neighbor, when he is enjoined to love me as himself, will answer exactly as I have done and will repel me for the same reasons. I hope he will not have the same objective grounds for doing so, but he will have the same idea as I have. Even so, the behavior of human beings shows differences, which ethics, disregarding the fact that such differences are determined, classifies as "good" or "bad." So long as these undeniable differences have not been removed, obedience to high ethical demands entails damage to the aims of civilization, for it puts a positive premium on being bad. One is irresistibly reminded of an incident in the French Chamber when capital punishment was being debated. A member had been passionately supporting its abolition and his speech was being received with tumultuous applause, when a voice from the hall called out: "Que messieurs les assassins commencent!"[6]

. . . Men are not gentle creatures who want to be loved, and who at the most can defend themselves if they are attacked; they are, on the contrary, creatures among whose instinctual endowments is to be reckoned a powerful share of aggressiveness. As a result, their neighbor is for them not only a potential helper or sexual object, but also someone who tempts them to satisfy their aggressiveness on him, to exploit his capacity for work without compensation, to use him sexually without his consent, to seize his possessions, to humiliate him, to cause him pain, to torture and to kill him. *Homo homini lupus.*[7] Who, in the face of all his experience of life and of history, will have the courage to dispute this assertion? As a rule this cruel aggressiveness waits for some provocation or puts itself at the service of some other purpose, whose goal might also have been reached by milder measures. In circumstances that are favorable to it, when the mental counter-forces which ordinarily inhibit it are out of action, it also manifests itself spontaneously and reveals man as a savage beast to whom consideration towards his own kind is something alien. Anyone who calls to mind the atrocities committed during the racial migrations or the invasions of the Huns, or by the people known as Mongols under Jenghiz Khan and Tamerlane, or at the capture of Jerusalem by the pious Crusaders, or even, indeed, the horrors of the recent World War—anyone who calls these things to mind will have to bow humbly before the truth of this view.

The existence of this inclination to aggression, which we can detect in ourselves and justly assume to be present in others, is the factor which disturbs our relations with our neighbor and

---

[6] ["It's the murderers who should make the first move."—TRANS.]
[7] ["Man is a wolf to man."—TRANS.]

which forces civilization into such a high expenditure [of energy]. In consequence of this primary mutual hostility of human beings, civilized society is perpetually threatened with disintegration. The interest of work in common would not hold it together; instinctual passions are stronger than reasonable interests. Civilization has to use its utmost efforts in order to set limits to man's aggressive instincts and to hold the manifestations of them in check by psychical reaction-formations. Hence, therefore, the use of methods intended to incite people into identifications and aim-inhibited relationships of love,[8] hence the restriction upon sexual life, and hence too the ideal's commandment to love one's neighbor as oneself—a commandment which is really justified by the fact that nothing else runs so strongly counter to the original nature of man. In spite of every effort, these endeavors of civilization have not so far achieved very much. It hopes to prevent the crudest excesses of brutal violence by itself assuming the right to use violence against criminals, but the law is not able to lay hold of the more cautious and refined manifestations of human aggressiveness. The time comes when each one of us has to give up as illusions the expectations which, in his youth, he pinned upon his fellowmen, and when he may learn how much difficulty and pain has been added to his life by their ill-will. At the same time, it would be unfair to reproach civilization with trying to eliminate strife and competition from human activity. These things are undoubtedly indispensable. But opposition is not necessarily enmity; it is merely misused and made an *occasion* for enmity.

*   *   *

What means does civilization employ in order to inhibit the

---

[8] [Non-sexual affection, including family attachment, friendship, and comradely feeling.]

aggressiveness which opposes it, to make it harmless, to get rid of it, perhaps? We have already become acquainted with a few of these methods, but not yet with the one that appears to be the most important. This we can study in the history of the development of the individual. What happens in him to render his desire for aggression innocuous? Something very remarkable, which we should never have guessed and which is nevertheless quite obvious. His aggressiveness is introjected, internalized; it is, in point of fact, sent back to where it came from—that is, it is directed towards his own ego. There it is taken over by a portion of the ego, which sets itself over against the rest of the ego as superego,[9] and which now, in the form of "conscience," is ready to put into action against the ego the same harsh aggressiveness that the ego would have liked to satisfy upon other, extraneous individuals. The tension between the harsh superego and the ego that is subjected to it, is called by us the sense of guilt; it expresses itself as a need for punishment. Civilization, therefore, obtains mastery over the individual's dangerous desire for aggression by weakening and disarming it and by setting up an agency within him to watch over it, like a garrison in a conquered city.

As to the origin of the sense of guilt, the analyst has different views from other psychologists; but even he does not find it easy to give an account of it. To begin with, if we ask how a person comes to have a sense of guilt, we arrive at an answer which cannot be disputed: a person feels guilty (devout people would say "sinful") when he has done something which he knows to be "bad." But then we notice how little this answer tells us. Perhaps, after some hesitation, we shall add that even when a person has not actually *done* the bad thing but has only recognized in himself an *intention* to do it, he may regard himself

---

[9] [An agency of the ego which holds up ideals for the ego to follow and chastises it when it fails to do so.]

as guilty; and the question then arises of why the intention is regarded as equal to the deed. Both cases, however, presuppose that one had already recognized that what is bad is reprehensible, is something that must not be carried out. How is this judgment arrived at? We may reject the existence of an original, as it were natural, capacity to distinguish good from bad. What is bad is often not at all what is injurious or dangerous to the ego; on the contrary, it may be something which is desirable and enjoyable to the ego. Here, therefore, there is an extraneous influence at work, and it is this that decides what is to be called good or bad. Since a person's own feelings would not have led him along this path, he must have had a motive for submitting to this extraneous influence. Such a motive is easily discovered in his helplessness and his dependence on other people, and it can best be designated as fear of loss of love. If he loses the love of another person upon whom he is dependent, he also ceases to be protected from a variety of dangers. Above all, he is exposed to the danger that this stronger person will show his superiority in the form of punishment. At the beginning, therefore, what is bad is whatever causes one to be threatened with loss of love. For fear of that loss, one must avoid it. This, too, is the reason why it makes little difference whether one has already done the bad thing or only intends to do it. In either case the danger only sets in if and when the authority discovers it, and in either case the authority would behave in the same way.

This state of mind is called a "bad conscience"; but actually it does not deserve this name, for at this stage the sense of guilt is clearly only a fear of loss of love, "social" anxiety. In small children it can never be anything else, but in many adults, too, it has only changed to the extent that the place of the father or the two parents is taken by the larger human community. Consequently, such people habitually allow themselves to do any bad thing which promises them enjoyment, so long as they are sure that the authority will not know anything about it or cannot blame them for it; they are afraid only of being found out. Present-day society has to reckon in general with this state of mind.

A great change takes place only when the authority is internalized through the establishment of a superego. The phenomena of conscience then reach a higher stage. Actually, it is not until now that we should speak of conscience or a sense of guilt. At this point, too, the fear of being found out comes to an end; the distinction, moreover, between doing something bad and wishing to do it disappears entirely, since nothing can be hidden from the superego, not even thoughts. It is true that the seriousness of the situation from a real point of view has passed away, for the new authority, the superego, has no motive that we know of for ill-treating the ego, with which it is intimately bound up; but genetic influence, which leads to the survival of what is past and has been surmounted, makes itself felt in the fact that fundamentally things remain as they were at the beginning. The superego torments the sinful ego with the same feeling of anxiety and is on the watch for opportunities of getting it punished by the external world.

At this second stage of development, the conscience exhibits a peculiarity which was absent from the first stage and which is no longer easy to account for. For the more virtuous a man is, the more severe and distrustful is its behavior, so that ultimately it is precisely those people who have carried saintliness furthest who reproach themselves with the worst sinfulness. This means that virtue forfeits some part of its promised reward; the docile and continent ego does not enjoy the trust of its mentor, and strives in vain, it would seem, to acquire it. The objection will at once be made that these difficulties are artificial ones, and it will be said that a stricter and more vigilant conscience is precisely the hallmark of a moral man. Moreover, when saints call themselves sinners, they are not so wrong, considering the temptations to instinctual satisfaction to which they are exposed in a specially high degree—since, as is well known, temptations are merely increased by constant frustration, whereas an occasional satisfaction of them causes them to diminish, at least for the time being. The field of ethics, which is so full of problems, presents us with another fact: namely that ill-luck—that is, external frustration—so greatly enhances the power of the conscience in the superego. As long as things go well with a man, his conscience is lenient and lets the ego do all sorts of things; but when misfortune befalls him, he searches his soul, acknowledges his sinfulness, heightens the demands of his conscience, imposes abstinences on himself and punishes himself with penances. Whole peoples have behaved in this way, and still do. This, however, is easily explained by the original infantile stage of conscience, which, as we see, is not given up after the introjection into the superego, but persists alongside of it and behind it. Fate is regarded as a substitute for the parental agency. If a man is unfortunate it means that he is no longer loved by this highest power; and, threatened by such a loss of love, he once more bows to the parental representative in his superego—a representative whom, in his days of good fortune, he was ready to neglect. This becomes especially clear where Fate is looked upon in the strictly religious sense of being nothing else than an expression of the Divine Will. The people of Israel had believed themselves to be the favorite child of God, and when the great Father caused misfortune after misfortune to rain down upon this people of his, they were never shaken in their belief in his relationship to them or questioned his power or righteousness. Instead, they produced the prophets, who held up their sinfulness before them; and out of their sense of guilt they created the over-strict commandments of their priestly religion. It is remarkable how differently a primitive man behaves. If he has met with a misfortune, he does not throw the blame on himself but on his fetish, which has obviously not done its duty, and he gives it a thrashing instead of punishing himself.

Thus we know of two origins of the sense of guilt: one arising from fear of an authority, and the other, later one, arising from fear of the superego. The first insists upon a renunciation of instinctual satisfactions; the second, as well as doing this, presses for punishment, since the continuance of the forbidden wishes cannot be concealed from the superego. We have also learned how the severity of the superego—the demands of conscience —is to be understood. It is simply a continuation of the severity of the external authority, to which it has succeeded and which it has in part replaced. We now see in what relationship the renunciation of instinct stands to the sense of guilt. Originally, renunciation of instinct was the result of fear of an external authority: one renounced one's satisfactions in order not to lose its love. If one has carried out this renunciation, one is, as it were, quits with the authority and no sense of guilt should remain. But with fear of the superego the case is different. Here, instinctual renunciation is not enough, for the wish persists and cannot be concealed from the superego. Thus, in spite of the renunciation that has been made, a sense of guilt comes about. . . . A threatened external unhappiness—loss of love and punishment on the part of the external authority—has been exchanged for a permanent internal unhappiness, for the tension of the sense of guilt.

*   *   *

For a wide variety of reasons, it is very far from my intention to express an opinion upon the value of human civilization. I have endeavored to guard myself against the enthusiastic prejudice which holds that our civilization is the most precious thing that we possess or could acquire and that its path will necessarily lead to heights of unimagined perfection. I can at least listen without indignation to the critic who is of the opinion that when one surveys the aims of cultural endeavor and the means it employs, one is bound to come to the conclusion that the whole effort is not worth the trouble, and that the outcome of it can only be a state of affairs which the individual will be unable to tolerate. My impartiality is made all the easier to me by my knowing very little about all these things. One thing only do I know for certain and that is that man's judgments of value follow directly his wishes for happiness—that, accordingly, they are an attempt to support his illusions with arguments. I should find it very understandable if someone were to point out the obligatory nature of the course of human civilization and were to say, for instance, that the tendencies to a restriction of sexual life or to the institution of a humanitarian ideal at the expense of natural selection were developmental trends which cannot be averted or turned aside and to which it is best for us to yield as though they were necessities of nature. I know, too, the objection that can be made against this, to the effect that in the history of mankind, trends such as these, which were considered unsurmountable, have often been thrown aside and replaced by other trends. Thus I have not the courage to rise up before my fellow-men as a prophet, and I bow to their reproach that I can offer them no consolation: for at bottom that is what they are all demanding—the wildest revolutionaries no less passionately than the most virtuous believers.

JEAN-JACQUES ROUSSEAU was born in 1712 in
Geneva, Switzerland. "My birth was the first of my
misfortunes," Rousseau mourned, meaning that his
mother had died after giving birth to him. Rousseau
ran away from home at sixteen and became a vagabond.
He wrote, "Never did I think so much, nor exist so
much, never was I so much myself as in the journeys I
made alone and on foot." During his wanderings,
Rousseau—a Calvinist—met Mme de Warens, a lay
proselytizer who converted him to Catholicism and later
seduced him. After their long affair ended in 1740,
Rousseau labored to make his name in Paris. He
invented a new system of musical notation; composed
an opera and a comedy; and served as secretary to
the French ambassador in Venice. But fame eluded him
until his essay "Discourse on the Sciences and the
Arts" won the Dijon Academy prize in 1750. Rousseau's philosophical writings were banned and the
author was exiled from France and Switzerland because
he and his work were thought to threaten existing
governments and religions. Rousseau also wrote novels,
a dictionary of music, and autobiographical prose before
he died in 1778.

A selection from *On the Social Contract,* translated by
Judith R. Masters and edited by Roger D. Masters.
Publisher: St. Martin's Press, Inc., 1978. Pages 46–76.

The Social Contract

BOOK I

I want to inquire whether there can be a legitimate and reliable rule of administration in the civil order, taking men as they are and laws as they can be. I shall try always to reconcile in this research what right permits with what interest prescribes, so that justice and utility are not at variance.

I start in without proving the importance of my subject. It will be asked if I am a prince or a legislator to write about politics. I reply that I am neither, and that is why I write about politics. If I were a prince or a legislator, I would not waste my time saying what has to be done. I would do it, or keep silent.

Born a citizen of a free State, and a member of the sovereign, the right to vote there is enough to impose on me the duty of learning about public affairs, no matter how feeble the influence of my voice may be. And I am happy, every time I meditate about governments, always to find in my research new reasons to love that of my country!

Subject of This First Book

Man was/is born free, and everywhere he is in chains. One who believes himself the master of others is nonetheless a greater slave than they. How did this change occur? I do not know. What can make it legitimate? I believe I can answer this question.

If I were to consider only force and the effect it produces, I would say that as long as a people is constrained to obey and does so, it does well; as soon as it can shake off the yoke and does so, it does even better. For in recovering its freedom by means of the same right used to steal it, either the people is justified in taking it back, or those who took it away were not justified in doing so. But the social order is a sacred right that serves as a basis for all the others. However, this right does not come from nature; it is therefore based on conventions. The problem is to know what these conventions are. Before coming to that, I should establish what I have just asserted.

On the First Societies

The most ancient of all societies, and the only natural one, is that of the family. Yet children remain bound to the father only as long as they need him for self-preservation. As soon as this need ceases, the natural bond dissolves. The children, exempt from the obedience they owed the father, and the father, exempt from the care he owed the children, all return equally to independence. If they continue to remain united, it is no longer naturally but voluntarily, and the family itself is maintained only by convention.

This common freedom is a consequence of man's nature. His first law is to attend to his own preservation, his first cares are those he owes himself; and as soon as he has reached the age of reason, as he alone is the judge of the proper means of preserving himself, he thus becomes his own master.

The family is therefore, so to speak, the prototype of political societies. The leader is like the father, the people are like the children; and since all are born equal and free, they only alienate their freedom for their utility. The entire difference is that in the family, the father's love for his children rewards him for the care he provides; whereas in the State, the pleasure of commanding substitutes for this love, which the leader does not have for his people. . . .

On the Right of the Strongest

The strongest is never strong enough to be the master forever unless he transforms his force into right and obedience into duty. This leads to the right of the strongest, a right that is in appearance taken ironically and in principle really established. But won't anyone ever explain this word to us? Force is a physical power. I do not see what morality can result from its effects. Yielding to force is an act of necessity, not of will. At most, it is an act of prudence. In what sense could it be a duty?

Let us suppose this alleged right for a moment. I say that what comes of it is nothing but inexplicable confusion. For as soon as force makes right, the effect changes along with the cause. Any force that overcomes the first one succeeds to its right. As soon as one can disobey without punishment, one can do so legitimately, and since the strongest is always right, the only thing to do is to make oneself the strongest. But what is a right that perishes when force ceases? If it is necessary to obey by force, one need not obey by duty, and if one is no longer forced to obey, one is no longer obligated to do so. It is apparent, then, that this word right adds nothing to force. It is meaningless here.

Obey those in power. If that means yield to force, the precept is good, but superfluous; I reply that it will never be violated. All power comes from God, I admit, but so does all illness. Does this mean it is forbidden to call the doctor? If a brigand takes me by surprise at the edge of a woods, must I not only give up my purse by force; am I obligated by conscience to give it even if I could keep it away? After all, the pistol he holds is also a power.

Let us agree, therefore, that might does not make right, and that one is only obligated to obey legitimate powers. Thus my original question still remains.

On Slavery

Since no man has any natural authority over his fellow man, and since force produces no right, there remain only conventions as the basis of all legitimate authority among men.

If a private individual, says Grotius, can alienate his freedom and enslave himself to a master, why can't a whole people alienate its freedom and subject itself to a king? There are many equivocal words in this that need explaining, but let us limit ourselves to the word *alienate*. To alienate is to give or to sell. Now a man who makes himself another's slave does not give himself, he sells himself, at the least for his subsistence. But why does a people sell itself? Far from furnishing the subsistence of his subjects, a king derives his own only from them, and according to Rabelais a king does not live cheaply. Do the subjects give their persons, then, on condition that their goods will be taken too? I do not see what remains for them to preserve.

It will be said that the despot guarantees civil tranquillity to his subjects. Perhaps so, but what have they gained if the wars that his ambition brings on them, if his insatiable greed, if the harassment of his ministers are a greater torment than their dissensions would be? What have they gained, if this tranquillity is one of their miseries? Life is tranquil in jail cells, too. Is that reason enough to like them? The Greeks lived tranquilly shut up in the Cyclop's cave as they awaited their turn to be devoured.

To say that a man gives himself gratuitously is to say something absurd and inconceivable. Such an act is illegitimate and null, if only because he who does so is not in his right mind. To say the same thing about an entire people is to suppose a people of madmen. Madness does not make right.

Even if everyone could alienate himself, he could not alienate his children. They are born men and free. Their freedom belongs to them; no one but themselves has a right to dispose of it. Before they have reached the age of reason, their father can, in their name, stipulate conditions for their preservation, for their well-being; but he cannot give them irrevocably and unconditionally, because such a gift is contrary to the ends of nature and exceeds the rights of paternity. For an arbitrary government to be legitimate, it would therefore be necessary for the people in each generation to be master of its acceptance or rejection. But then this government would no longer be arbitrary.

To renounce one's freedom is to renounce one's status as a man, the rights of humanity and even its duties. There is no possible compensation for anyone who renounces everything. Such a renunciation is incompatible with the nature of man, and taking away all his freedom of will is taking away all morality from his actions. Finally, it is a vain and contradictory convention to stipulate absolute authority on one side and on the other unlimited obedience. Isn't it clear that one is in no way engaged toward a person from whom one has the right to demand everything, and doesn't this condition alone—without equivalent and without exchange—entail the nullification of the act? For what right would my slave have against me, since all he has belongs to me, and his right being mine, my right against myself is a meaningless word?

Grotius and others derive from war another origin of the alleged right of slavery. As the victor has the right to kill the vanquished, according to them, the latter can buy back his life at the cost of his freedom—a convention all the more legitimate in that it is profitable for both of them.

But it is clear that this alleged right to kill the vanquished in no way results from the state of war. Men are not naturally enemies, if only because when living in their original independence, they do not have sufficiently stable relationships among themselves to constitute either the state of peace or the state of war. It is the relationship between things, not between men, that constitutes war; and as the state of war cannot arise from simple, personal relations, but only from proprietary relations, private war between one man and another can exist neither in the state of nature, where there is no stable property, nor in the social state, where everything is under the authority of the laws. . . .

War is not, therefore, a relation between man and man, but between State and State, in which private individuals are enemies only by accident, not as men, nor even as citizens, but as soldiers; not as members of the homeland but as its defenders. Finally, each State can have only other States, and not men, as enemies, since no true relationship can be established between things of differing natures.

This principle even conforms with the established maxims of all ages and with the constant practice of all civilized peoples. Declarations of war are not so much warnings to those in power as to their subjects. The foreigner—whether he be king, private individual, or people—who robs, kills, or imprisons subjects without declaring war on the prince, is not an enemy, but a brigand. Even in the midst of war, a just prince may well seize everything in an enemy country that belongs to the public, but he respects the person and goods of private individuals. He respects rights on which his own are based. The end of war being the destruction of the enemy State, one has the right to kill its defenders as long as they are armed. But as soon as they lay down their arms and surrender, since they cease to be enemies or instruments of the enemy, they become simply men once again, and one no longer has a right to their lives. Sometimes it is possible to kill the State without killing a single one of its members. War confers no right that is not necessary to its end. These principles are not those of Grotius; they are not based on the authority of poets, but are derived from the nature of things, and are based on reason.

With regard to the right of conquest, it has no basis other than the law of the strongest. If war does not give the victor the right to massacre the vanquished peoples, this right he does not have cannot establish the right to enslave them. One only has the right to kill the enemy when he cannot be made a slave. The right to make him a slave does not come, then, from the right to kill him. It is therefore an iniquitous exchange to make him buy his life, over which one has no right, at the cost of his freedom. By establishing the right of life and death on the right of slavery, and the right of slavery on the right of life and death, isn't it clear that one falls into a vicious circle?

Even assuming this terrible right to kill everyone, I say that a man enslaved in war or a conquered people is in no way obligated toward his master, except to obey for as long as he is forced to do so. In taking the equivalent of his life, the victor has not spared it; rather than to kill him purposelessly, he has killed him usefully. Therefore, far from the victor having acquired any authority over him in addition to force, the state of war subsists between them as before; their relation itself is its effect, and the customs of the right of war suppose that there has not been a peace treaty. They made a convention, true; but that convention, far from destroying the state of war, assumes its continuation.

Thus, from every vantage point, the right of slavery is null, not merely because it is illegitimate, but because it is absurd and meaningless. These words *slavery* and *right* are contradictory; they are mutually exclusive. Whether it is said by one man to another or by a man to a people, the following speech will always be equally senseless: *I make a convention with you that is entirely at your expense and entirely for my benefit; that I shall observe for as long as I want, and that you shall observe for as long as I want.*

That It Is Always Necessary to Go Back to a First Convention

Even if I were to grant everything I have thus far refuted, the proponents of despotism would be no better off. There will always be a great difference between subjugating a multitude and governing a society. If scattered men, however many there may be, are successively enslaved by one individual, I see only a master and slaves; I do not see a people and its leader. It is an aggregation, if you wish, but not an association. It has neither public good nor body politic. That man, even if he had enslaved half the world, is nothing but a private individual. His interest, separate from that of the others, is still nothing but a private interest. If this same man dies, thereafter his empire is left scattered and without bonds, just as an oak tree disintegrates and falls into a heap of ashes after fire has consumed it.

A people, says Grotius, can give itself to a king. According to Grotius, a people is therefore a people before it gives itself to a king. This gift itself is a civil act; it presupposes a public deliberation. Therefore, before examining the act by which a people elects a king, it would be well to examine the act by which a people becomes a people. For this act, being necessarily prior to the other, is the true basis of society. . . .

On the Social Compact

I assume that men have reached the point where obstacles to their self-preservation in the state of nature prevail by their resistance over the forces each individual can use to maintain himself in that state. Then that primitive state can no longer subsist and the human race would perish if it did not change its way of life.

Now since men cannot engender new forces, but merely unite and direct existing ones, they have no other means of self-preservation except to form, by aggregation, a sum of forces that can prevail over the resistance; set them to work by a single motivation; and make them act in concert.

This sum of forces can arise only from the cooperation of many. But since each man's force and freedom are the primary instruments of his self-preservation, how is he to engage them without harming himself and without neglecting the cares he owes to himself? In the context of my subject, this difficulty can be stated in these terms:

"Find a form of association that defends and protects the person and goods of each associate with all the common force, and by means of which each one, uniting with all, nevertheless obeys only himself and remains as free as before." This is the fundamental problem which is solved by the social contract.

The clauses of this contract are so completely determined by the nature of the act that the slightest modification would render them null and void. So that although they may never have been formally pronounced, they are everywhere the same, everywhere tacitly accepted and recognized, until the social compact is violated, at which point each man recovers his original rights and resumes his natural freedom, thereby losing the conventional freedom for which he renounced it.

Properly understood, all of these clauses come down to a single one, namely the total alienation of each associate, with all his rights, to the whole community. For first of all, since each one gives his entire self, the condition is equal for everyone, and since the condition is equal for everyone, no one has an interest in making it burdensome for the others.

Furthermore, as the alienation is made without reservation, the union is as perfect as it can be, and no associate has anything further to claim. For if some rights were left to private individuals, there would be no common superior who could judge between them and the public. Each man being his own judge on some point would soon claim to be so on all; the state of nature would subsist and the association would necessarily become tyrannical or ineffectual.

Finally, as each gives himself to all, he gives himself to no one; and since there is no associate over whom one does not acquire the same right one grants him over oneself, one gains the equivalent of everything one loses, and more force to preserve what one has.

If, then, everything that is not the essence of the social compact is set aside, one will find that it can be reduced to the following terms: *Each of us puts his person and all his power in common under the supreme direction of the general will; and in a body we receive each member as an indivisible part of the whole.*

Instantly, in place of the private person of each contracting party, this act of association produces a moral and collective body, composed of as many members as there are voices in the assembly, which receives from this same act its unity, its common *self,* its life, and its will. This public person, formed thus by the union of all the others, formerly took the name *City,* and now takes that of *Republic* or *body politic,* which its members call *State* when it is passive, *Sovereign* when active, *Power* when comparing it to similar bodies. As for the associates, they collectively take the name *People;* and individually are called *Citizens* as participants in the sovereign authority, and *Subjects* as subject to the laws of the State. But these terms are often mixed up and mistaken for one another. It is enough to know how to distinguish them when they are used with complete precision.

On the Sovereign

This formula shows that the act of association includes a reciprocal engagement between the public and private individuals, and that each individual, contracting with himself so to speak, finds that he is doubly engaged, namely toward private individuals as a member of the sovereign and toward the sovereign as a member of the State. But the maxim of civil right that no one can be held responsible for engagements toward himself cannot be applied here, because there is a great difference between being obligated to oneself, or to a whole of which one is a part.

It must further be noted that the public deliberation that can obligate all of the subjects to the sovereign—due to the two different relationships in which each of them is considered—cannot for the opposite reason obligate the sovereign toward itself; and that consequently it is contrary to the nature of the body politic for the sovereign to impose on itself a law it cannot break. Since the sovereign can only be considered in a single relationship, it is then in the situation of a private individual contracting with himself. It is apparent from this that there is not, nor can there be, any kind of fundamental law that is obligatory for the body of the people, not even the social contract. This does not mean that this body cannot perfectly well enter an engagement toward another with respect to things that do not violate this contract. For with reference to the foreigner, it becomes a simple being or individual.

But the body politic or the sovereign, deriving its being solely from the sanctity of the contract, can never obligate itself, even toward another, to do anything that violates that original act, such as to alienate some part of itself or to subject itself to another sovereign. To violate the act by which it exists would be to destroy itself, and whatever is nothing, produces nothing.

As soon as this multitude is thus united in a body, one cannot harm one of the members without attacking the body, and it is even less possible to harm the body without the members feeling the effects. Thus duty and interest equally obligate the two contracting parties to mutual assistance, and the same men should seek to combine in this double relationship all the advantages that are dependent on it.

Now the sovereign, formed solely by the private individuals composing it, does not and cannot have any interest contrary to theirs. Consequently, the sovereign power has no need of a guarantee toward the subjects, because it is impossible for the body ever to want to harm all its members, and we shall see later that it cannot harm any one of them as an individual. The sovereign, by the sole fact of being, is always what it ought to be.

But the same is not true of the subjects in relation to the sovereign, which, despite the common interest, would have no guarantee of the subjects' engagements if it did not find ways to be assured of their fidelity.

Indeed, each individual can, as a man, have a private will contrary to or differing from the general will he has as a citizen. His private interest can speak to him quite differently from the common interest. His absolute and naturally independent existence can bring him to view what he owes the common cause as a free contribution, the loss of which will harm others less than its payment burdens him. And considering the moral person of the State as an imaginary being because it is not a man, he might wish to enjoy the rights of the citizen without wanting to fulfill the duties of a subject, an injustice whose spread would cause the ruin of the body politic.

Therefore, in order for the social compact not to be an ineffectual formula, it tacitly includes the following engagement, which alone can give force to the others: that whoever refuses to obey the general will shall be constrained to do so by the entire body; which means only that he will be forced to be free. For this is the condition that, by giving each citizen to the homeland, guarantees him against all personal dependence; a condition that creates the ingenuity and functioning of the political machine, and alone gives legitimacy to civil engagements without which it would be absurd, tyrannical, and subject to the most enormous abuses.

On the Civil State

This passage from the state of nature to the civil state produces a remarkable change in man, by substituting justice for instinct in his behavior and giving his actions the morality they previously lacked. Only then, when the voice of duty replaces physical impulse and right replaces appetite, does man, who until that time only considered himself, find himself forced to act upon other principles and to consult his reason before heeding his inclinations. Although in this state he deprives himself of several advantages given him by nature, he gains such great ones, his faculties are exercised and developed, his ideas broadened, his feelings ennobled, and his whole soul elevated to such a point that if the abuses of this new condition did not often degrade him beneath the condition he left, he ought ceaselessly to bless the happy moment that tore him away from it forever, and that changed him from a stupid, limited animal into an intelligent being and a man.

Let us reduce the pros and cons to easily compared terms. What man loses by the social contract is his natural freedom and an unlimited right to everything that tempts him and that he can get; what he gains is civil freedom and the proprietorship of everything he possesses. In order not to be mistaken about these compensations, one must distinguish carefully between natural freedom, which is limited only by the force of the individual, and civil freedom, which is limited by the general will; and between possession, which is only the effect of force or the right of the first occupant, and property, which can only be based on a positive title.

To the foregoing acquisitions of the civil state could be added moral freedom, which alone makes man truly the master of himself. For the impulse of appetite alone is slavery, and obedience to the law one has prescribed for oneself is freedom. But I have already said too much about this topic, and the philo-

sophic meaning of the word *freedom* is not my subject here. . . .
I shall end . . . this book with a comment that ought to serve
as the basis of the whole social system. It is that rather than
destroying natural equality, the fundamental compact on the
contrary substitutes a moral and legitimate equality for whatever
physical inequality nature may have placed between men, and
that although they may be unequal in force or in genius, they
all become equal through convention and by right.

Book II

That Sovereignty Is Inalienable

The first and most important consequence of the principles established above is that the general will alone can guide the forces
of the State according to the end for which it was instituted,
which is the common good. For if the opposition of private
interests made the establishment of societies necessary, it is the
agreement of these same interests that made it possible. It is
what these different interests have in common that forms the
social bond, and if there were not some point at which all the
interests are in agreement, no society could exist. Now it is
uniquely on the basis of this common interest that society ought
to be governed.

I say, therefore, that sovereignty, being only the exercise of
the general will, can never be alienated, and that the sovereign,
which is only a collective being, can only be represented by itself.
Power can perfectly well be transferred, but not will.

Indeed, though it is not impossible for a private will to agree
with the general will on a given point, it is impossible, at least,
for this agreement to be lasting and unchanging. For the private
will tends by its nature toward preferences, and the general will
toward equality. It is even more impossible for there to be a
guarantee of this agreement even should it always exist. It would
not be the result of art, but of chance. The sovereign may well

say, "I currently want what a particular man wants, or at least what he says he wants." But it cannot say, "What that man will want tomorrow, I shall still want," since it is absurd for the will to tie itself down for the future and since no will can consent to anything that is contrary to the good of the being that wills. Therefore, if the people promises simply to obey, it dissolves itself by that act; it loses the status of a people. The moment there is a master, there is no longer a sovereign, and from then on the body politic is destroyed.

This is not to say that the commands of leaders cannot pass for expressions of the general will, as long as the sovereign, being free to oppose them, does not do so. In such a case, one ought to presume the consent of the people from universal silence. This will be explained at greater length. . . .

Whether the General Will Can Err

From the foregoing it follows that the general will is always right and always tends toward the public utility. But it does not follow that the people's deliberations always have the same rectitude. One always wants what is good for oneself, but one does not always see it. The people is never corrupted, but it is often fooled, and only then does it appear to want what is bad.

There is often a great difference between the will of all and the general will. The latter considers only the common interest; the former considers private interest, and is only a sum of private wills. But take away from these same wills the pluses and minuses that cancel each other out, and the remaining sum of the differences is the general will.

If, when an adequately informed people deliberates, the citizens were to have no communication among themselves, the general will would always result from the large number of small differences, and the deliberation would always be good. But when factions, partial associations at the expense of the whole, are formed, the will of each of these associations becomes general with reference to its members and particular with reference to the State. One can say, then, that there are no longer as many voters as there are men, but merely as many as there are associations. The differences become less numerous and produce a result that is less general. Finally, when one of these associations is so big that it prevails over all the others, the result is no longer a sum of small differences, but a single difference. Then there is no longer a general will, and the opinion that prevails is merely a private opinion.

In order for the general will to be well expressed, it is therefore important that there be no partial society in the State, and that each citizen give only his own opinion. Such was the unique and sublime system instituted by the great Lycurgus. If there are partial societies, their number must be multiplied and their inequality prevented, as was done by Solon, Numa, and Servius. These precautions are the only valid means of ensuring that the general will is always enlightened and that the people is not deceived.

On the Limits of the Sovereign Power

If the State or the City is only a moral person whose life consists in the union of its members, and if the most important of its concerns is that of its own preservation, it must have a universal, compulsory force to move and arrange each part in the manner best suited to the whole. Just as nature gives each man absolute power over all his members, the social compact gives the body politic absolute power over all its members, and it is this same power directed by the general will, which as I have said bears the name sovereignty.

But in addition to the public person, we have to consider the private persons who compose it and whose life and freedom are naturally independent of it. It is a matter, then, of making a clear distinction between the respective rights of the citizens and the sovereign, and between the duties that the former have to fulfill as subjects and the natural rights to which they are entitled as men.

It is agreed that each person alienates through the social compact only that part of his power, goods, and freedom whose use matters to the community; but it must also be agreed that the sovereign alone is the judge of what matters.

A citizen owes the State all the services he can render it as soon as the sovereign requests them. But the sovereign, for its part, cannot impose on the subjects any burden that is useless to the community. It cannot even will to do so, for under the law of reason nothing is done without a cause, any more than under the law of nature.

The engagements that bind us to the social body are obligatory only because they are mutual, and their nature is such that in fulfilling them one cannot work for someone else without also working for oneself. Why is the general will always right and why do all constantly want the happiness of each, if not because there is no one who does not apply this word *each* to himself, and does not think of himself as he votes for all? Which proves that the equality of right, and the concept of justice it produces, are derived from each man's preference for himself and consequently from the nature of man; that the general will, to be truly such, should be general in its object as well as in its essence; that it should come from all to apply to all; and that it loses its natural rectitude when it is directed toward any individual, determinate object. Because then, judging what is foreign to us, we have no true principle of equity to guide us. . . .

It should be understood from this that what generalizes the will is not so much the number of votes as the common interest that unites them, because in this institution everyone necessarily subjects himself to the conditions he imposes on others, an admirable agreement between interest and justice which confers on common deliberations a quality of equity that vanishes in the discussion of private matters, for want of a common interest that unites and identifies the rule of the judge with that of the party.

However one traces the principle, one always reaches the same conclusion, namely that the social compact established an equality between the citizens such that they all engage themselves under the same conditions and should all benefit from the same rights. Thus by the very nature of the compact, every act of sovereignty, which is to say every authentic act of the general will, obligates or favors all citizens equally, so that the sovereign knows only the nation as a body and makes no distinctions between any of those who compose it. What really is an act of sovereignty then? It is not a convention between a superior and an inferior, but a convention between the body and each of its members. A convention that is legitimate because it has the social contract as a basis; equitable, because it is common to all; useful, because it can have no other object than the general good; and solid, because it has the public force and the supreme power as guarantee. As long as subjects are subordinated only to such conventions, they do not obey anyone, but solely their own will; and to ask how far the respective rights of the sovereign and of citizens extend is to ask how far the latter can engage themselves to one another, each to all and all to each.

It is apparent from this that the sovereign power, albeit entirely absolute, entirely sacred, and entirely inviolable, does not and cannot exceed the limits of the general conventions, and that every man can fully dispose of the part of his goods and freedom that has been left to him by these conventions. So that the sovereign never has the right to burden one subject more than another, because then the matter becomes individual, and its power is no longer competent.

Once these distinctions are acknowledged, it is so false that the social contract involves any true renunciation on the part of private individuals that their situation, by the effect of this contract, is actually preferable to what it was beforehand; and instead of an alienation, they have only exchanged to their advantage an uncertain, precarious mode of existence for another that is better and safer; natural independence for freedom; the power to harm others for their personal safety; and their force, which others could overcome, for a right which the social union makes invincible. Their life itself, which they have dedicated to the State, is constantly protected by it; and when they risk it for the State's defense, what are they then doing except to give back to the State what they have received from it? What are they doing that they did not do more often and with greater danger in the state of nature, when waging inevitable fights they defend at the risk of their life that which preserves it for them? It is true that everyone has to fight, if need be, for the homeland, but also no one ever has to fight for himself. Don't we still gain by risking, for something that gives us security, a part of what we would have to risk for ourselves as soon as our security is taken away?

On the Right of Life and Death

It is asked how private individuals who have no right to dispose of their own lives can transfer to the sovereign a right they do not have. This question appears hard to resolve only because it is badly put. Every man has a right to risk his own life in order to preserve it. Has it ever been said that someone who jumps out of a window to escape a fire is guilty of suicide? Has this crime ever even been imputed to someone who dies in a storm, although he was aware of the danger when he set off?

The social treaty has the preservation of the contracting parties as its end. Whoever wants the end also wants the means, and these means are inseparable from some risks, even from some losses. Whoever wants to preserve his life at the expense of others should also give it up for them when necessary. Now the citizen is no longer judge of the risk to which the law wills that he be exposed, and when the prince has said to him, "It is expedient for the State that you should die," he ought to die. Because it is only under this condition that he has lived in safety up to that point, and because his life is no longer only a favor of nature, but a conditional gift of the State.

The death penalty inflicted on criminals can be considered from approximately the same point of view: it is in order not to be the victim of a murderer that a person consents to die if he becomes one. Under this treaty, far from disposing of one's own life, one only thinks of guaranteeing it; and it cannot be presumed that any of the contracting parties is at that time planning to have himself hanged.

Besides, every offender who attacks the social right becomes through his crimes a rebel and traitor to his homeland; he ceases to be one of its members by violating its laws, and he even wages war against it. Then the State's preservation is incompatible with his own, so one of the two must perish; and when the guilty man is put to death, it is less as a citizen than as an enemy. The proceedings and judgment are the proofs and declaration that he has broken the social treaty, and consequently is no longer a member of the State. Now as he had acknowledged himself to be such, at the very least by his residence, he ought to be removed from it by exile as a violator of the compact or by death as a public enemy. For such an enemy is not a moral person but a man, and in this case the right of war is to kill the vanquished.

But it will be said that the condemnation of a criminal is a particular act. Agreed—hence this condemnation is not to be made by the sovereign. It is a right the sovereign can confer without itself being able to exercise it. All of my ideas fit together, but I can hardly present them simultaneously.

Moreover, frequent corporal punishment is always a sign of weakness or laziness in the government. There is no wicked man who could not be made good for something. One only has the right to put to death, even as an example, someone who cannot be preserved without danger.

With regard to the right to pardon, or to exempt a guilty man from the penalty prescribed by the law and pronounced by the judge, this belongs only to one who is above the judge and the law—which is to say, to the sovereign. Yet its right in this matter is not very clear and the cases in which it is applied are very rare. In a well-governed State, there are few punishments, not because may pardons are given, but because there are few criminals. When the State declines, a high number of crimes guarantees their impunity. Under the Roman Republic, the senate and consuls never tried to pardon. The people itself did not do so, although it sometimes revoked its own judgment. Frequent pardons indicate that crimes will soon have no further need of them, and everyone sees where that leads. But I feel that my heart murmurs and holds back my pen. Let these questions be discussed by the just man who has never transgressed and who never needed pardon himself.

On Law

Through the social compact we have given the body politic existence and life; the issue now is to give it movement and will through legislation. For the original act which forms and unites this body does not thereby determine anything about what it should do to preserve itself.

Whatever is good and in accordance with order is so by the nature of things, independently of human conventions. All justice comes from God; He alone is its source. But if we knew how to receive it from on high, we would need neither government nor laws. There is without doubt a universal justice emanating from reason alone; but to be acknowledged among us, this justice must be reciprocal. Considering things from a human point of view, the laws of justice are ineffectual among men for want of a natural sanction. They merely benefit the wicked man and harm the just, when the latter observes them toward everyone while no one observes them toward him. Therefore, there must be conventions and laws to combine rights with duties and to bring justice back to its object. In the state of nature where everything is in common, I owe nothing to those to whom I have promised nothing; I recognize as belonging to someone else only what is useless to me. It is not the same in the civil state where all rights are fixed by the law.

But what is a law after all? As long as people are satisfied to attach only metaphysical ideas to this word, they will continue to reason without understanding each other, and when they have stated what a law of nature is, they will not thereby have a better idea of what a law of the State is.

I have already said that there is no general will concerning a particular object. Indeed, that particular object is either within the State or outside of the State. If it is outside of the State, a will that is foreign to it is not general in relation to it. And if within the State, that object is part of it. Then a relation between the whole and its parts is formed which makes of them two separate entities, one of which is the part and the other of which is the whole minus that part. But the whole minus a part is not the whole, and for as long as this relationship lasts, there is no whole, but rather two unequal parts. It follows from this that the will of one of them is no longer general in relation to the other.

But when the entire people enacts something concerning the entire people, it considers only itself, and if a relationship is formed then, it is between the whole object viewed in one way and the whole object viewed in another, without any division of the whole. Then the subject matter of the enactment is general like the will that enacts. It is this act that I call a law.

When I say that the object of the law is always general, I mean that the law considers the subjects as a body and actions in the abstract, never a man as an individual or a particular action. Thus the law can very well enact that there will be privileges, but it cannot confer them on anyone by name. The law can create several classes of citizens, and even designate the qualities determining who has a right to these classes, but it cannot name the specific people to be admitted to them. It can establish a royal government and hereditary succession, but it cannot elect a king or name a royal family. In short, any function that relates to an individual object does not belong to the legislative power.

Given this idea, one sees immediately that it is no longer necessary to ask who should make laws, since they are acts of the general will; nor whether the prince is above the laws, since he is a member of the State; nor whether the law can be unjust, since no one is unjust toward himself; nor how one is free yet subject to the laws, since they merely record our wills.

Furthermore, one sees that since the law combines the universality of the will and that of the object, what any man, whoever he may be, orders on his own authority is not a law. Whatever is ordered even by the sovereign concerning a particular object is not a law either, but rather a decree; nor is it an act of sovereignty, but of magistracy.

I therefore call every State ruled by laws a republic, whatever the form of administration may be, for then alone the public interest governs and the commonwealth really exists.

*   *   *

On the Various Systems of Legislation

If one seeks to define precisely what constitutes the greatest good of all, which ought to be the end of every system of legislation, one will find that it comes down to these two principal objects: *freedom* and *equality*. Freedom because all private dependence is that much force subtracted from the body of the State; equality because freedom cannot last without it.

I have already said what civil freedom is. With regard to equality, this word must not be understood to mean that degrees of power and wealth should be exactly the same, but rather that with regard to power, it should be incapable of all violence and never exerted except by virtue of status and the laws; and with regard to wealth, no citizen should be so opulent that he can buy another, and none so poor that he is constrained to sell himself. This presumes moderation in goods and influence on the part of the upper classes and moderation in avarice and covetousness on the part of the lower classes.

This equality is said to be a speculative fantasy that cannot exist in practice. But if abuse is inevitable, does it follow that it must not at least be regulated? It is precisely because the force of things always tends to destroy equality that the force of legislation should always tend to maintain it.

But these general objects of all good institutions should be modified in each country according to the relationships that arise as much from the local situation as from the character of the inhabitants, and it is on the basis of these relationships that each people must be assigned a particular system of institutions that is the best, not perhaps in itself, but for the State for which it is intended. For example, is the soil unprofitable and barren, or the country too small for its inhabitants? Turn to industry and the arts, the products of which can be exchanged for the foodstuffs you lack. On the contrary, do you inhabit rich plains and fertile hillsides? Do you lack inhabitants on good terrain? Apply all your efforts to agriculture, which multiplies men, and chase out the arts, which would merely complete the country's depopulation by concentrating its small number of inhabitants in a few locations. Do you inhabit extensive, convenient shores? Cover the sea with ships; cultivate commerce and navigation.

You will have a brilliant and brief existence. Is your coast merely a place where the sea meets almost inaccessible rocks? Remain barbarians and fish eaters; you will live more peacefully, better perhaps, and surely more happily. In short, apart from the maxims common to all, each people contains within itself some cause that organizes it in a particular manner and renders its legislation appropriate for it alone. Thus the Hebrews long ago and the Arabs recently have had religion as their principal object; the Athenians, letters; Carthage and Tyre, commerce; Rhodes, navigation; Sparta, war; and Rome, virtue. The author of *The Spirit of Laws*[1] has given large numbers of examples of the art by which the legislator directs the institutions toward each of these objects.

The constitution of a State is made truly solid and enduring when matters of expediency are so well satisfied that natural relationships and the laws always agree on the same points, and the latter only secure, accompany, and rectify, so to speak, the former. But if the legislator makes a mistake about his objective and adopts a different principle from the one arising from the nature of things—whether one tends toward servitude and the other toward freedom, one toward wealth and the other toward population growth, or one toward peace and the other toward conquest—the laws will imperceptibly weaken, the constitution will be altered, and the State will not cease being agitated until it is either destroyed or changed, and invincible nature has regained its dominion.

---

[1] [Montesquieu.]